A Vision for the U.S. Forest Service

A Vision for the U.S. Forest Service

Goals for Its Next Century

c.1

Edited by
Roger A. Sedjo

Resources for the Future
Washington, DC

Printed in the United States of America

An RFF Press book
Published by Resources for the Future
1616 P Street, NW, Washington, DC 20036–1400
www.rff.org

Library of Congress Cataloging-in-Publication Data

A vision for the U.S. Forest Service : goals for its next century / edited by Roger A. Sedjo.
p. cm.
Papers presented at a conference held April 1999 in Washington D.C.
ISBN 1–891853–02–3
1. United States. Forest Service—Management—Congresses. 2. Forest policy—United States—Congresses. 3. Forest reserves—Management—Government policy—United States—Congresses. I. Sedjo, Roger A.
SD565 .V57 2000
333.75'0973—dc21 00–032333

f e d c b a

The paper in this book meets the guidelines for permanence and durability of the Committee on Production Guidelines for Book Longevity of the Council on Library Resources.

This book was typeset in Palatino by Amie Jackowski and Betsy Kulamer. It was copyedited by Pamela Angulo. The cover was designed by Debra Naylor Design.

About Resources for the Future and RFF Press

Founded in 1952, Resources for the Future (RFF) contributes to environmental and natural resource policymaking worldwide by performing independent social science research.

RFF pioneered the application of economics as a tool to develop more effective policy about the use and conservation of natural resources. Its scholars continue to employ social science methods to analyze critical issues concerning pollution control, energy policy, land and water use, hazardous waste, climate change, biodiversity, and the environmental challenges of developing countries.

RFF Press supports the mission of RFF by publishing book-length works that present a broad range of approaches to the study of natural resources and the environment. Its authors and editors include RFF staff, researchers from the larger academic and policy communities, and journalists. Audiences for RFF publications include all of the participants in the policymaking process—scholars, the media, advocacy groups, NGOs, professionals in business and government, and the general public.

Contents

Preface

The U.S. Department of Agriculture Forest Service has a long and honorable history. However, as the organization approaches its 100th birthday, it finds itself in deep trouble. The legislation that defines its work—the Resources Planning Act, as amended by the National Forest Management Act—are products of the 1970s.

The consensus that once supported the "multiple" activities of the Forest Service has largely dissipated. Although some observers argue that existing legislation is adequate for the effective functioning of the Forest Service today, others disagree. Clearly, even if the legislation is legally adequate, the Forest Service lacks a well-articulated, unambiguous mission.

The regulations under which the Forest Service functions today were issued in 1982. Several attempts to update these regulations have been unsuccessful. Even as I write this preface, another effort to revise the Forest Service regulations is under way in conjunction with the recent report of the second Committee of Scientists, or COS (of which I am a member). This committee was created by the secretary of agriculture to examine problems of forest planning and offer some assistance in revising the regulations. The COS report has received mixed reviews at best, and the success of this effort remains in doubt.

In this volume, the authors address two primary issues: why the Forest Service faces difficulty, and how it might overcome those difficulties and regain a well-respected position. Certainly, no one is suggesting that the agency return to the old model. The world has changed, and an effectively operating Forest Service must adapt to the new world that has evolved.

Earlier versions of most of the chapters of this book were first presented as papers at a national conference hosted by Resources for the Future, entitled "A Vision for the Future: Where Should the Forest Service Be in 2009, and How Might It Get There?" held April 29–30, 1999, in Washington, DC. The conference brought together some of the best scholars and practitioners who have spent much of their lives thinking about and wrestling with issues that involve the management of public resources, particularly the public forests. Consensus among the authors is that the Forest Service faces serious—some would say potentially lethal—challenges. The authors agree that times have changed since the Forest Service was founded, that the American people have differing ideas about what ought to be the major purposes of Forest Service management, and that Forest Service personnel no longer share a common vision for managing the national forests.

The possible sources of the problems that face today's Forest Service are many: the serious divisions that exist today among the American people in their view of forest management objectives; a political process that politicizes questions of public lands management; the (perhaps) inherent contentiousness of the American people, or a least our political/legal system, which leads to never-ending appeals and litigation; the past and present failure of the Forest Service to provide leadership; the inherent flaws in large bureaucratic organizations that rely on the planning process rather than the market; the lack of a balance among the Forest Service's constituencies; and perhaps many others.

Finding a way to resolve these issues, however, is not an easy task. Can the Forest Service be reformed? While some argue for decentralization and greater responsiveness to markets, others maintain that the Forest Service was specifically created to be insulated from the market—in effect, to be responsive to political pressures instead of market pressures. Was the Forest Service really more effective earlier in its history, when it was more removed from the political process? Furthermore, it appears that the faith the public once held in scientific and technical expertise is largely absent today. Also, it is now recognized that science and technology identify the management methods required for achieving objectives but do not determine the end objectives.

Many people deserve thanks for their contributions to both this volume and the conference that generated these papers. Former Deputy Associate Chief of the Forest Service Jerry Sesco first developed the idea of a critical, multifaceted examination of the Forest Service. I commend the Forest Service and Chief Mike Dombeck for their willingness to provide funding for an activity that was certain to be at least partly critical of many of its efforts. I wish to thank Paul R. Portney, president of Resources for the Future, and Michael A. Toman, director of RFF's Energy and Natural Resources Division, for their support. The Pinchot Institute also provided assistance in organizing the conference. Of course, I wish to thank the authors whose papers are published in this volume and Resources for the Future for its support—financial and moral—of this project.

Finally, Marion Clawson deserves thanks for his continuing inspiration. He surely would have applauded this attempt to examine and assess the Forest Service and its management of the National Forest System, a task to which he devoted a considerable portion of his professional life. Marion had an intense interest in public lands issues, particularly as related to the National Forest System and the Forest Service. Most of his professional efforts in the latter part of his career were focused on examining these issues. This volume is dedicated to his memory and in his honor.

ROGER A. SEDJO
Resources for the Future

In memory of
Marion Clawson

1

Marion Clawson and America's Forests

A Lifetime of Commitment

Roger A. Sedjo

For twenty years, Marion Clawson and I were colleagues at Resources for the Future (RFF). I first met Marion when I began my career at RFF in the late 1970s. At that time, he was more than seventy years old and "technically" retired. I knew of Marion by virtue of his professional reputation and also knew that he was still among the most active and productive researchers at RFF. Of course, Marion never really retired.

During my first decade at RFF, Marion arrived in the office early every morning. He produced about one book every two years. At any point in time, Marion had one book at the publisher, was in the process of writing another, and was thinking about a third. Over the next ten years, Marion published five more books, the final one (in 1987) his memoirs. He continued to write occasional papers and came to his office regularly until his death in the spring of 1998 at the age of ninety-two. He remained a close colleague and friend to the end.

Marion's work was of academic interest but also policy-relevant for many of the pressing public lands issues of the time. Marion garnered a respect from the policy community that researchers rarely attain, perhaps because early in his career, he had been closely involved with the same kinds of management issues he later addressed from a research perspective. Although Marion did not go out of his way to enter the political fray,

ROGER A. SEDJO is a senior fellow at Resources for the Future and director of RFF's Forest Economics and Policy Program.

when he did become involved, he typically drew the attention of the policymakers and the agencies.

Many of Marion's forays into forestry cast light on inefficient public management practices. Using the agricultural model developed in his early training, he concluded that inputs were justified only if they generated outputs that were of greater value. He criticized the Forest Service for spending large amounts of money to improve forests on marginal sites, where the returns on investment were low. He also criticized the Forest Service for not maintaining an appropriately high level of harvest on public lands. He was not alone in this criticism. In his 1983 book, Marion noted with approval that "the Office of Management and Budget (OMB) and the General Accounting Office (GAO) have repeatedly pressured the Forest Service into making larger timber sales in order to accelerate the rate of harvest of old-growth timber" (Clawson 1983, 82). After all, in the early 1980s, with the absence of a widespread perspective that old-growth stands were inherently valuable, most economists maintained that rational, efficient multiple-use management should include the felling of most of these trees for timber use and the subsequent regeneration of lands for additional forest harvests. Old-growth values could be captured in parks and protected areas that were already set aside. The issues of the time involved how rapidly old growth ought to be liquidated, how best to regenerate the forest, and how to accomplish these goals cost-effectively. Marion willingly entered this debate.

Marion was a remarkable man. His lifetime spanned almost the entire twentieth century. His interests included a host of important areas and topics, most if not all of which had to do with the interaction between human beings and natural resources.

THE EARLY YEARS

Marion Clawson grew up in Nevada. A product of western America, his earliest memories were of life on the small ranch where his father was a rancher and miner. Marion's interest in ranching was reflected in his academic work. He received a Bachelor of Science degree in agriculture in 1926 and a Master of Science in agricultural economics in 1929, both from the University of Nevada.

In essence, Marion had two full careers. His first was as a civil servant, initially in the U.S. Department of Agriculture. He started out doing agricultural research out west and eventually moved to Washington, DC. During World War II, he earned a Ph.D. in economics from Harvard University. He joined the Bureau of Land Management (BLM) in 1947 and became its director in 1948. In 1953, he was fired as director of the BLM by the incoming Republican administration, a feat of which he was quite proud. This termination provided him the perfect opportunity to start a second career.

A FORAY INTO FORESTRY

After spending two years in Israel as a member of a foreign economic advisory staff, Marion began his second career as a researcher with RFF. He had published three books prior to his RFF tenure. The first two, published in 1947 and 1950, were about agriculture. In his third book, *Uncle Sam's Acres* (Clawson 1951), he shifted his focus to the topic that would become the subject of the majority of his work from that point: public lands. His first RFF book, *The Federal Lands: Their Use and Management* (Clawson and Held 1957), was about land. During the early part of his RFF career (roughly 1955 through the early 1970s), Marion's work covered a host of topics, including agriculture, soil conservation, and urban land policy. However, the principal focuses of his work were land issues and outdoor recreation; his contributions to these areas were substantial.

Marion's first major involvement in forestry was as one of several authors of the *Report of the President's Advisory Panel on Timber and the Environment* (President's Advisory Panel 1973). Among other things, the report called for "the Federal agencies concerned with forests to prepare a comprehensive nationwide program of forest development and timber supply...." Years later, however, Marion opined that the report had had a significant effect on the development of the National Forest Management Act (NFMA), which was passed in 1976. Although the report was attributed to the entire advisory panel, Marion wrote much of it himself. He had been approached by Paul McCraken, chairman of the President's Council of Economic Advisors, to participate on the panel. Although Marion protested that he had little experience in forestry, McCraken had persisted. At the time it was published, the report was not overwhelmingly influential. It was delivered to President Richard Nixon as the Watergate hearings were heating up and got lost in the general chaos that ensued.

This experience marked a turning point in Marion's already diverse and illustrious career. After that panel report, Marion went on to write eleven more books, eight of which were on forestry. In the 1980s, RFF President Emery Castle confided to me that after the panel report, he had occasionally suggested to Marion that he return to some of his earlier interests. "But," Castle said, "Marion just doesn't seem to be interested in anything but forestry."

Perhaps Marion's most influential book on forestry is *Forests for Whom and for What?* (Clawson 1975). In this book, Marion addressed public forest land issues ranging from timber production to timberland withdrawals from harvest, examining these issues from the perspective of an economist interested in both the commodity and nonmarket outputs of the forest. He addressed not only issues of economic efficiency but also cultural and social acceptability and the consequences of forest production on income distribution. In many respects, this modest book was an early primer on forest economics, forest issues, and forest policy. Although it was published almost three decades ago, *Forests for Whom and for What?* still is frequently cited in the literature.

Much of Marion's work on forests focused on the public forest land and the National Forest System. His books in this area include *Forest Policy for the Future* (Clawson 1974), *The Economics of National Forest Management* (Clawson 1976a), and *The Federal Lands Revisited* (Clawson 1983). Marion's concerns involved raising questions as much as providing answers—for example, "to what uses lands should be put" and "how best to manage lands for those ends." In this context, the mix of federal, state, and private lands is important. Also important are questions such as how many federal dollars should be spent on various desired uses and how these lands could be managed efficiently to meet these ends. And finally, if land disposal or acquisition by the federal estate were desired, how could this best be accomplished?

The peak of Marion's influence on forestry probably came during the late 1970s and early 1980s. In his final book on forestry, *The Federal Lands Revisited* (Clawson 1983), Marion again addressed the management of the federal forests. Two very influential articles in the journal *Science* preceded this book. In the first article, "The National Forests—A Great National Asset Is Poorly Managed and Unproductive" (Clawson 1976b), Marion argued that the National Forest Service devoted too many resources to poor lands and too few resources to high-productivity sites. In the second article, "Forests in the Long Sweep of American History" (Clawson 1979), Marion showed how the nation's forests had recovered, far beyond what even the most optimistic analysts had anticipated, from earlier logging and land-clearing abuses. His argument was that the American forests were in far better condition than commonly supposed, largely because of their natural resiliency, which he felt was consistently underestimated. A healthy and dynamic U.S. forest system was recently "rediscovered" (Wernick and others 1998).

In the early 1980s, with the absence of a widespread perspective that old-growth stands were inherently valuable, most economists maintained that rational, efficient multiple-use management should include the felling of most of these trees for timber use and the subsequent regeneration of the lands for additional forest harvests. Only toward the latter part of the 1980s did the fate of the spotted owl seriously enter the discussion of old-growth forest values. Quickly, the ongoing debate was substantially modified via the constraints of the Endangered Species Act—the viability provision of the federal regulations—and large areas of old-growth forest were set aside for spotted owl conservation. Marion's view was that if one of the roles of the Forest Service was to produce timber—and he argued that this mandate went back to the Forest Reserve Act of 1891 (see Clawson 1983, 72) and straight through the NFMA of 1976—then they ought to do it efficiently.

Marion tended to view the NFMA with cautious approval. Like John Krutilla, another giant in the field of forest economics whose career overlapped Marion's at RFF, he broadly interpreted the 1976 act as requiring management consistent with economic maximization (Clawson 1983, 181). However, Marion was uncomfortable with the planning language, noting that "while requiring such balancing of costs and benefits in the planning process," it "only implicitly (not explicitly) requires that the resultant plans shall govern the actual administrative actions of the Forest Service" (Clawson 1983, 181). This problem was rediscovered in a GAO report (U.S. GAO 1997), which noted the prevalent absence of implementation, and also was noted in the Committee of Scientists' report (USDA 1999).

One of Marion's unique intellectual traits was the ability to look at a well-recognized problem from a slightly different perspective. Time and time again, he would challenge conventional wisdom by taking an unconventional view of a problem and coming up with an unorthodox but useful perspective. For example, in *The Economics of U.S. Nonindustrial Private Forests* (Clawson 1978), Marion demonstrated that much of the difference in productivity between the national forests and the poorly regarded nonindustrial private forests (NIPFs) was due to location and age, not management. He noted that the NIPFs are disproportionately located in regions whose climates and other characteristics contribute to modest biological growth. Once adjustments are made for these considerations, the NIPFs perform comparably to the national forests. These findings were not always well accepted, and even I was drawn into some of the rancorous debates—both formal and informal—that ensued. However, Marion's perspective generated a surge in papers and research on these important forests, which constitute 58% of the total forested area in the United States.

Marion's views might best be characterized as those of Pinchot-type conservationism, in contrast to the Muir-type preservationism that is ascendant today. Having grown up in Nevada, the son of a miner and rancher who just barely eked out a living from the earth (Clawson 1987), Marion viewed resources as something to be used, but used sensibly. His training as an economist served to add emphasis and rigor to his concerns regarding the importance of the efficient use of resources. Resources were to be used—sustainably.

He appreciated the nontimber values of the forest and supported the idea of multiple-use management. As noted, much of Marion's early work at RFF was focused on outdoor recreation and its valuation. Furthermore, he understood that a major rationale of public ownership of forest lands was based on the desire for multiple-use outputs, many of which were not valued in the market and therefore were unlikely to be produced in appropriate quantities by the market (Clawson 1983, 136–42).

Marion also recognized the value and role of parks and wilderness. Again, to him, the issue was not whether to establish parks and wilderness but how much and what kinds of natural areas should be established as such. On several occasions, his writings addressed the question of how much forest should be retained and how much and what kinds of lands ought to be made off-limits to harvesting (Clawson 1975, 7, 8, 158, and 159). Even though Marion's belief in nontimber values and the importance of wilderness to society was strong, he maintained the belief that an important role of the National Forest Service, as explicitly stated in the Forest Reserve Act and the Organic Act, was to provide for future timber requirements (Clawson 1983, 72–77). During the latter decades of the twentieth century, he believed that it was appropriate for us to use those resources.

Although Marion held very strong opinions, he had the admirable ability to rethink his positions. He was a conservationist whom I would characterize as a "New Deal" Democrat. He saw an important role for the federal lands and devoted much of his life's work to studying and analyzing how they might best fulfill that role. Thus, it came as somewhat of a surprise to me that he seriously considered some of the arguments for the privatization of the public lands that were presented early in the Reagan administration.

In his 1983 book, Marion analyzed the pros and cons of privatization of parts of the federal estate. From my private conversations with Marion, I surmised that some of the reluctant enthusiasm that he exhibited for privatization reflected, in part, his frustration with much of federal management. Also, he recognized that at various times in its history, the federal government sometimes focused on acquiring lands and other times focused on land disposal. It is worth noting that all of the alternatives he suggested and examined (that is, retention in federal ownership with strenuous efforts to improve their management, transfer to the states, privatization of all or major parts of the National Forest Service, transfer to public or mixed public–private corporations, and long-term leasing) were assessed in terms of their ability to provide more efficient land management (Clawson 1983, 177).

Marion was not a consistent "friend" of the Forest Service; he often criticized it for inefficiency. However, neither was he a consistent environmentalist. A Pinchot-type conservationist, Marion never developed an appreciation for the people he called "preservationists," by which he seemed to mean those who were primarily interested in "locking up resources." Like other economists, he was concerned with balancing alternative and sometimes conflicting values, and he tended to reject views that focused exclusively on a single value—whether they came from preservationists or timber barons. Logging had its place in the forest, as did other outputs and values.

Marion also had the ability to recognize his mistakes. Many times, he acknowledged how badly he had misinterpreted the public forest experience of the 1950s. The increasing federal harvests of that period—and the associated increased revenues—had suggested to him that the Forest Service could actually cover the costs of its operations and perhaps would generate net revenues to the U.S. Treasury through its timber management (see Clawson and Held 1957).

At the beginning of the nineteenth century, Pinchot had promised that the Forest Service could and would generate net revenues from forestry. This potential profit was used to justify the view of the Forest Service as a different and special type of agency: It was to provide outputs as well as maintain and protect the forest. This perspective provided the rationale for locating the Forest Service in the U.S. Department of Agriculture, whose responsibility also included production and investments in the land. However, Pinchot was never able to deliver on his promise to make forestry pay. Similarly, Marion's expectations did not anticipate the increased interest in environmental issues and wilderness that would take place in the 1960s and beyond. These interests created pressure for lower timber harvests and greater forest land set-asides, a trend that was inconsistent with his expectation of increasing net revenues.

A PIONEERING LEGACY

Marion left behind a considerable professional legacy. He was a pioneer, one of the founding fathers of the discipline called resource and environmental economics. He left a legacy of more than thirty professional books and hundreds of published papers. His publications contributed to the development of the field and are partly responsible for where it is today, for the choice of problems it examines and how it considers issues. Marion recognized that resources were limited and believed that they should be used efficiently by both private and public users.

He also understood the existence of values that were, in essence, economic values beyond those recognized and traded in the market. He contributed to a method that addresses such issues. Even today, the "travel cost" method, whereby travel and other expenditures are used to estimate individual and collective demand curves for recreation, is frequently referred to as the Clawson Demand Curve. Yet, he never could be accused of being in an "ivory tower." Methods were developed intended to be applied to problems in the real world. Throughout his career, Marion was concerned with primarily real-world problems.

Marion saw a role for governments in explicitly managing for the values that were not well-represented in markets. They included forest resources, which generate a mix of outputs, both market and nonmarket. He also saw that bureaucracies often tended to lose sight of the purposes for which they were created and often were not truly concerned about the "wise use" of resources.

He brought candor and honesty to his field. In his professional career and, as best I can judge, in his personal life, Marion "called it as he saw it." In his profession, he was forthright, sometimes almost blunt, but always gentle. He never was mean-spirited. He had a civility in his professional demeanor that we would do well to emulate today.

THE FINAL CHAPTER

Although Marion became less involved in the daily goings-on at RFF in his later years, he was scheduled to attend my Forestry Economics and Policy Program Advisory Committee meeting in May 1998. He had not missed one of these meetings in the previous twenty years! Sadly, we lost him one month before that meeting.

During his last several years, Marion had faithfully visited the office for a couple of hours every Wednesday morning—to pick up his mail, have a letter typed, or chat briefly with the fellow researchers and staff. Shortly before his passing, he shared with me a suggestion from his son Pat, who was concerned that at the age of ninety-two, Marion still drove through the morning rush-hour traffic to the office. Pat had told him, "Dad, there are taxis that you can take to and from work." Marion, feigning irritation, had replied, "Pat, I know there are taxis, and when I need a taxi, I will call a taxi."

He never did call a taxi. And that is the way I always will remember him.

ACKNOWLEDGMENTS

First and foremost, I applaud Marion Clawson's contributions to forestry and forest economics, which are reflected in the volumes of forestry literature he created. Second, thanks to Pamela Jagger for useful comments and critiques on earlier drafts of this document. Any errors that remain are mine. Finally, I acknowledge the USDA Forest Service for providing partial funds for the development of this chapter, this entire book, and the related national conference, "A Vision for the Forest Service: Where Should the Forest Service Be in 2009 and How Might It Get There?" held in April 1999.

REFERENCES

Clawson, Marion. 1951. *Uncle Sam's Acres.* New York: Dodd Meade Publishers.

Clawson, Marion, ed. 1974. *Forest Policy for the Future: Conflict, Compromise, Consensus.* Washington, DC: Resources for the Future.

Clawson, Marion. 1975. *Forests for Whom and for What?* Baltimore, MD: Johns Hopkins Univ. Press for Resources for the Future.

———. 1976a. *The Economics of National Forest Management.* Washington, DC: Resources for the Future.

———. 1976b. The National Forests: A Great National Asset Is Poorly Managed and Unproductive. *Science* 191(4227): 762–7.

———. 1978. *The Economics of U.S. Nonindustrial Private Forests.* Washington, DC: Resources for the Future.

———. 1979. Forests in the Long Sweep of American History. *Science* 204(4398): 1168–74.

———. 1983. *The Federal Lands Revisited.* Washington, DC: Resources for the Future.

———. 1987. *From Sage Brush to Sage.* Washington, DC: ANA Publications.

Clawson, M., and R. B. Held. 1957. *The Federal Lands: Their Use and Management.* Baltimore, MD: Johns Hopkins University Press for Resources for the Future.

President's Advisory Panel. 1973. *Report of the President's Advisory Panel on Timber and the Environment.* Washington, DC: U.S. Government Printing Office, 541.

USDA (U.S. Department of Agriculture) Committee of Scientists. 1999. *Sustaining the People's Lands: Recommendations for Stewardship of the National Forests and Grasslands into the Next Century.* Washington, DC: U.S. Department of Agriculture.

U.S. GAO (General Accounting Office). 1997. *Forest Service Decision-Making: A Framework for Improving Performance.* GAO/RCED-97-71. Washington, DC: U.S. General Accounting Office.

Wernick, I. K., P. E. Waggoner, and J. H. Ausubel. 1998. Searching for Leverage to Conserve Forests: Industrial Ecology of Wood Products in the U.S. *Journal of Industrial Ecology* 1(3): 125–45.

2

What Now?

From a Former Chief of the Forest Service

Jack Ward Thomas

The Forest Service finds itself at a crossroads as we approach the 100th anniversary of its founding. My purpose in this chapter is to make some suggestions, based on experience, for possible modifications in law, operations, and budget that might smooth the transition to the next century.

I admittedly begin and end with a strong bias. I believe that the Forest Service—warts and all—is the best conservation organization in the world. The people of the past and present Forest Service have made it so. I came to the agency thirty-three years ago after ten years with a state wildlife agency because I simply wanted to be part of the Forest Service—part of something bigger than myself and an agency that set standards for the world.

I did not find perfection in the Forest Service. There were many squabbles as the times and the circumstances, knowledge, and desires of the American people changed. I was there when the emphasis shifted to intensive resource extraction and when it shifted yet again to focus on fish and wildlife, recreation, and water. Those changes reached a crescendo, and the Forest Service was caught with one foot rooted firmly in the past and the other tentatively venturing into the future.

The Forest Service's place in the future is no longer tentative. Change has come, and there is no going back. That much is clear. But the next step is less obvious—and a matter of intense debate. I hope that my experience and insight can help, just a bit, to light the path.

JACK WARD THOMAS is the Boone and Crockett Professor at the School of Forestry, University of Montana, and a former chief of the USDA Forest Service.

Some of these ideas were on the agenda when I left my position as chief of the Forest Service. I am appreciative that Chief Mike Dombeck has followed through on a few of them and added some new twists that I agree with and support. I refrain from discussing my disagreements here; those are for the chief's ears only. The Forest Service can have and needs only one chief at a time, and my tenure has ended.

TOO MANY LAWS? TIME FOR REVIEW

The plethora of laws that affect Forest Service management, particularly of national forests, has at best made management activities increasingly expensive, uncertain, unpredictable, contentious, unwieldy, and unlikely to take place. Examination of these laws and pursuant regulations in terms of their compatibility, especially when complicated by applicable case law, reveals serious problems with contradictions, overlaps in authority, and certain conflicts. These problems are exacerbated by differing missions for different agencies and the power struggles between regulatory and management agencies as each, in good faith, struggles to achieve its mission.

I have heard the combination of these laws referred to as a "hodgepodge," a "crazy quilt," and "a cloak of many colors." When I came to the chief's job, my superiors assured me there was no problem with the laws, and exacerbating problems were billed as an unwillingness of prior administrations to comply with these laws. At the time, I thought that interpretation was wrong. By the time I left the job, I *knew* it was wrong.

But who could argue with the intent of each of those laws when considered individually? And who could argue that national forest management does not become more complex and problematic as these laws are applied in combination? The confusion is then magnified as the playing field is constantly reconstituted by new case law.

It is time—past time—to take a careful look at the laws that influence the management of the national forests (and perhaps all public lands) and to develop legislation that will clarify missions, reconcile conflicts, and define authorities. The mood of almost vicious partisanship that so dominates Congress at this juncture—most notably, in the committees and subcommittees that deal with Forest Service matters—essentially precludes the likelihood of any improvement evolving directly through that channel. Where there is no will, there is no way. The Left dares not offend its hard-core environmental constituencies (who are pleased with the evolving situation); the Right plays to those who profit from resource extraction from public lands and looks back to yesterday with longing and the desire to recreate the past. The situation is such that Congress can neither lead nor follow, and, as politicians, they cannot get out of the way. So, how do we move toward a better situation than that which exists today?

One time-honored solution to handling such a hot potato is to name a large bipartisan commission, usually composed of big names with an interest. The real work, however, is done by a staff. The one certain result is the *appearance* that something is being done.

Maybe it is time again for a Public Land Law Review Commission—but with a twist. Maybe, for once, such a commission might be composed of real experts of proven competence with no particular ax to grind. The key to success would be to pick the right leader and allow that leader to put together a suitable team (with approval reserved to elected officials) and to develop the required budget. The team members should clearly understand the task at hand and have a fixed time to complete it. After appropriate analysis, including public hearings, the ultimate outcome would be draft legislation—perhaps in several forms.

WHAT IS THE FOREST SERVICE'S MISSION?

I probably spent more time in front of congressional committees in a shorter period than any of my predecessors, and my successor has likely broken my record. After the Republicans took over the Senate and House in the 1994 elections, the increasingly frequent hearings tended to display more and more acrimony—and produced little more than that.

Many times, I told those committees what I believe to be the crux of the debate over the national forests: The problem is not that the Forest Service does not have a clear and overriding mission but that the mission is not clearly spelled out in law and is not universally acknowledged. And the Forest Service can do little or nothing about it. The mission has simply evolved, with no open blessing from Congress or the administration.

The overriding policy for the management of the national forests is the preservation of biodiversity. That objective (policy) will be achieved first. That objective is closely followed by the assurance of water quality and then, if possible, some production of goods and services. In carrying out that policy, it is very difficult for the Forest Service to satisfy critics in Congress from the states with substantial federal lands who want to know why the Forest Service is not carrying out the mission. They define that mission as producing a "nondeclining even flow" of timber for their constituents who depend on federal timber, whether directly as employees or tax-paying industries, or indirectly in the form of receipts from resource exploitation to county treasuries. The same can be said of grazing and mineral extraction.

This situation has evolved as the direct result of the Endangered Species Act (ESA) of 1973 and the regulations issued by the Forest Service pursuant to the National Forest Management Act (NFMA) of 1976. The purpose of the ESA is "the preservation of ecosystems upon which threatened and endangered species depend." The clause in the regulations issued pursuant to NFMA says that "all native and desired nonnative vertebrates will be maintained in viable numbers well-distributed in the planning area." Oddly, this regulation is far more stringent in protecting species than the ESA is.

The period since the passage of the ESA has produced an ever-expanding list of plant and animal species considered threatened or endangered, and many more are likely to come. Disproportionate numbers of these species are found in national forests, in my opinion, not because the situation there is worse than the norm but because conditions are relatively better than on other ownerships. All of these species require special consideration in management.

With every such listing, the Forest Service gains a management partner with veto authority over proposed management actions. In the case of terrestrial species or aquatic life that is not connected with the oceans, the management partner is the Fish and Wildlife Service (U.S. Department of the Interior). In the case of anadromous species (that is, those that breed in fresh water and then go to the ocean), the National Marine Fisheries Service (U.S. Department of Commerce) is the partner. These partners are not equal. They have the duty to develop recovery plans and see that such plans are executed. That duty entails veto power over proposed actions.

Couple this with the requirements of the above-described clause in the Forest Service's planning regulations and the consequences of a number of federal court rulings that have interpreted those regulations literally. Then, consider the resultant effect on management actions. It should be crystal clear that retaining biodiversity is the overriding mission of national forest management.

How that set of circumstances has evolved bothers me—a lot. The overriding attention to biodiversity does not bother me so much as an appropriate policy or a mission. In fact, as a biologist, I support the mission, but with considerable reservation as to how management direction has evolved from regulatory agencies. It troubles me that this mission has simply evolved out of a series of laws and pursuant regulations, court cases, and policy direction. This evolved mission should be ratified—or rejected—by Congress and the administration. If it is determined to be the Forest Service's overriding mission, then so be it. If Congress disagrees, then it has a duty to clarify the situation. At least the committees in Congress should acknowledge this situation.

Frankly, I doubt that we will see any such clarification in the near term. Clarification would require a collective nerve that I do not believe exists. It is far easier to mollify constituents on the extremes of the preservation/exploitation debate by leaving the Forest Service in the middle to absorb the slings and arrows. But I will say this to my friends in Congress from the public land states and to those in the business of extracting and processing natural resources from the public's lands: If you expect anything other than a constant decline in the availability of goods and services from the national forests resulting from this de facto mission, you are indeed dreamers of what was, not what is and what will be.

When we think about the mission for the Forest Service and the management of the national forests, we should consider some facts:

- The population of the United States is less than 6% of that of the world.
- The population of the United States consumes more than 25% of the timber resources exploited over the entire Earth.
- The population of the United States continues to increase and will likely grow another 40–60% by 2050.
- The U.S. per capita consumption of wood and wood fiber is the highest in the world—and is increasing.
- The U.S. per capita income is increasing and is expected to continue to increase.
- The United States has the largest, best-trained cadre of natural resource managers in the world.
- The United States has the greatest capability in the world to train natural resources professionals.
- The policy of the U.S. government is to increase consumption of both goods and services.
- The U.S. Congress has emasculated international forestry programs that would assist other nations in the practice of sustainable forestry as they meet the demands of American consumers.
- The annual timber harvest from the U.S. national forests has declined from approximately twelve billion board feet per year to much less than four billion board feet and continues to decline.
- Because the wood yield from the U.S. national forests is well under 5% of current national consumption, some groups are pushing for a "zero cut."

These facts lead—or should lead—to nagging questions. Where does the wood for the world's greatest consumer come from over the long haul? The answer seems to be "elsewhere"—wherever that is. We are losing acreage from the national timber base at a steady rate as a result of development and subdivision. As the ownerships get smaller, they are less and less likely to be timber producers. Industrial forest lands are owned by corporations that are in business to make money for their stockholders, pure and simple. Such is the essence of capitalism: maximization of profit for owners. So, in a sense, any land-holding entity is simultaneously a land speculator. It seems likely that when the land is more valuable to satisfy that essence of the American dream—a place in the country—than it is for growing wood or some other higher and better use, the land will pass from the category of industrial forest land to nonindustrial forest land. Then, it will be turned into smaller and smaller ownerships.

After the national forests are out of the production of wood and wood fiber, is it likely that forestry on private lands will face increasing constraints imposed for environmental reasons and by higher costs? Quite probably. Then where will we get our wood? Elsewhere? What are the consequences of going elsewhere to meet our growing demands for wood? Much of the supply will, in my opinion, come from the poorer parts of the United States and the world, which have the fewest environmental constraints, the most people in need of work, and the biggest desire for American dollars.

Are there moral questions here? Do we, with our increasing wealth, simply say, "To hell with elsewhere"? No other place in the world has a better capability to practice sustainable, multiple-use natural resource management than the United States. Does that mean I defend all that has gone before? No. But thousands of practicing natural resources professionals have learned much and are learning more each day. We have learned from our successes and our failures—much like all professionals. If we can't practice sustainable forestry, nobody can.

Before we rely on elsewhere, we need to consider the moral aspects of that decision.

PROBLEMS

Micromanagement

By micromanagement, I mean the increasingly detailed instruction given to the Forest Service, at all levels, by political appointees of the administration in power, individuals in Congress, and Congress in general.

As examples, consider the following scenarios, which show how the constant pressures of micromanagement can counter good order, discipline, efficiency, effectiveness, and coherent operations:

- An assistant to the secretary of agriculture, without explanation or discussion with the chief, orders—or passes down orders—that a timber sale be withdrawn.
- Congress directs that two ranger districts cannot be combined.

- Congress allocates research dollars to a particular university scientist who holds favor with a member of the Appropriations Committee.
- The undersecretary of agriculture pushes the chief to promote favored individuals to particular jobs.
- The undersecretary rejects the chief's decisions on job selections.
- The secretary orders removal of particular senior executives, with no reasons given.
- The undersecretary refuses to allow the Forest Service, as was customary, to put forward a clearly identified Forest Service request on budget, thereby preventing Congress and the American people from knowing the chief's opinion relating to the budget.
- The Congress passes and the President signs the Salvage Rider without consultation with the Forest Service.
- The undersecretary orders the placement of selected congressional staffers in the civil service.
- Political appointees and members of Congress interfere in assessments, planning, and selection of planning alternatives.
- Powerful members of Congress make threats in clear efforts to intimidate Forest Service personnel.

As micromanagement—from the administration and the Congress—increases, things become more chaotic. This is particularly true when the White House is controlled by one party and the Congress by the other, and the two parties view Forest Service operations in a different light. The Forest Service gets caught squarely in the middle. The people of the Forest Service look to the chief for leadership and direction, and when they see the political machinations undercutting their vision of the chief—whoever he or she may be—they feel rudderless and dispirited. Much of the élan of the Forest Service is predicated on the crucial internal mythology that surrounds the professionalism of the agency as personified in the chief. It does not mean that the employees cannot change. It means that esprit de corps and pride are attributes that have made the Forest Service a special agency and a special place to work. Maintaining pride in achievement is essential. That esprit de corps is being dangerously eroded by increased micromanagement. Change? Yes. Micromanagement as a mechanism to make change? No.

This trend can be addressed two ways. The first is to place the Forest Service (and perhaps all land management agencies) under a quasi-public organization. The organization would operate under a board of directors appointed to ensure overlapping terms (for continuity) and bipartisan composition. The chief, selected from the ranks of the Forest Service, would be appointed for a set term to begin at the midpoint between presidential elections. [Note: I consider the current chief, Mike Dombeck, to be a professional from the ranks and tried and tested while being "on loan" to the Bureau of Land Management.]

The second, weaker alternative is to have the chief report directly to the secretary of agriculture by eliminating two layers (perhaps three—I was never sure) of intermediate political appointees. During my tenure in the Forest Service, I was personally acquainted with and dealt closely with several undersecretaries and assistant secretaries. Some knew next to nothing about natural resource management (particularly management of forested ecosystems), and some knew—or thought they knew—a lot. Some oversaw the Forest Service with a light hand, and others practiced almost daily internal involvement. These appointees should stick strictly with policy development and leave execution to the Forest Service. If policy is not appropriately pursued, then the chief should be replaced. An undersecretary who confuses his or her position with that of the chief is the ultimate threat to good order.

Over the years, undersecretaries have come and gone without much notice or much effect on Forest Service operations. More recently, we have seen two notable exceptions. One undersecretary appointed by Ronald Reagan pushed the Forest Service for a timber cut level far in excess of what seemed to many Forest Service professionals as either reasonable or sustainable, and the execution of those cut levels produced a backlash—and appropriately so. Another appointed by Bill Clinton pushed the Forest Service far in the opposite direction, producing a backlash in public land states and in committees in Congress. The whiplash effect produced by these powerful and dedicated men (each dedicated to a very different vision) has thrown the Forest Service and its various constituencies into confusion and a snapping, snarling dogfight.

The Forest Service's path no longer seems a smooth and evolutionary journey; it seems uncertain as to direction and even purpose. Watching Forest Service managers reminds me a bit of a scene from *The Wizard of Oz* movie, when the journey of Dorothy and company along the Yellow Brick Road to the Emerald City leads them through the forest. They look into the shadows and surmise that there lurk "lions and tigers and bears." As they walk on, they see specters in the forest and call out, "Lions and tigers and bears, oh my!" Feeding on their fears, they repeat, "LIONS and TIGERS and BEARS, OH MY!" louder and louder, and begin to run.

Now, the Forest Service line officers facing appeals, lawsuits, mixed directions, regulatory agencies, distressed community leaders, members of Congress, and micromanagement have to contend with lions and tigers and bears aplenty. These specters grow more vivid and make the Forest Service more and more cautious and less and less prone to action.

The chief should be clearly responsible for the Forest Service. When undersecretaries confined themselves to the policy arena, confusion and consternation were relatively minimal. As these political appointees have more and more confused themselves with the chief, operations have become more chaotic, and morale among the seasoned troops has taken a downward turn. Having the chief report directly to the secretary of agriculture has the disadvantage of not having a political shield between the chief and the secretary, thereby drawing the chief closer to the political arena. I believe this risk is reasonable—one clearly supported in history. The required direct contact with the chief would, I believe, tend to improve the secretary's interest in and appreciation for the Forest Service. I wonder what the first chief, Gifford Pinchot, would have been able to achieve if he had operated with two layers of political appointees between him and the secretary.

I am increasingly convinced that it is essentially impossible to manage natural resources with a hundred-year vision on the basis of a two-year election cycle. I also am convinced that the chief of the Forest Service should be a foremost proponent and spokesman for good natural resource management. To have a person in the chief's chair who is limited by the politics of organizational circumstances is a lamentable waste of talent and experience.

The Forest Service was once the can-do agency and a source of leadership for conservation efforts. Such attributes were the attributes of the people of the Forest Service, not of the chief, per se. In my opinion, the "genetic code" for such remains embedded in the people of the Forest Service. Some means of letting these genes express themselves is sorely needed, for the nation, for the people who are directly affected by Forest Service actions, and for the people of the Forest Service. That release may well rest in the inclusion of the chief of the Forest Service in policy-setting circles and not submerged two bureaucratic layers below the secretary of agriculture.

Planning Regulations Issued Pursuant to NFMA

When the sponsor of the NFMA, Senator Hubert Humphrey, stood on the Senate floor and said something to the effect that, "We have taken the management of the national forests out of the hands of the courts and placed it in the hands of the professionals," he could not have been more wrong. Planning proved to be much more expensive and has taken much more time than anticipated. It was expected to produce a consensus on the management of individual national forests but instead caused polarization. It was trumpeted as a way to increase stability and predictability, but both have decreased.

Congress expected planning, I believe, to be conducted by agency professionals—bottom to top—without coercion from political appointees.

Increasingly, that has not been the case. Many of the plans that emerged from the first round of planning included timber projections that were simply unrealistic, in my opinion, because of unrelenting pressure from the undersecretary of agriculture who had authority over the Forest Service at the time.

Then the Forest Service, in its best can-do style, set out to make it so. It was quickly apparent that there was a disconnect between plans and budgets at the forest level. The budgets for timber operations and roading were typically funded at requested levels (sometimes above requested levels), and other aspects of the plan (fish and wildlife, recreation, monitoring, and watershed) were funded at a fraction of requested amounts. Because the budget is *the operative policy* document, the Forest Service resolutely followed the budget instructions. The cumulative effect was inevitable and increasing conflict with environmental laws, regulatory agencies, and the rapidly growing environmental community.

The planning operations, because of appeals and lawsuits, became increasingly sensitive to the requirements for being "suit proof" or "appeal proof." This requirement lengthened the process and resulted in page after page of gobbledygook that only a lawyer or a dedicated technowonk could appreciate or understand. The meetings with the public increased in number and duration as months stretched to several years. As time marched on, all but the zealots and the hired guns dropped out of the process.

From this battleground rose the "conflict industry"—the hired guns, the gladiators, the warriors. The "firms," ensconced on the extremes of protection and exploitation, grew to include chief executive officers, technical experts, lawyers, fundraisers, publishers, press officers, field coordinators, and others.

Many of the problems arose from the regulations issued pursuant to NFMA. I believe the authors of NFMA considered that these regulations should, could, and would be tweaked on a regular basis as experience accumulated. And this tweaking, or full revision, would be handled by Forest Service professionals with cursory review by political appointees. The result? Any revision at all has proven impossible to achieve over twenty-three years. This continuing stalemate has resulted from political decisions above the chief's level, usually in the form of a political decision to withhold issuance of final regulations until after an election. Then, as a result of vagaries of political fortune, a change in power forced reconsideration of the proposed regulations. The planning regulations somehow have evolved from planning guidance to a political document. As failure after failure to produce new regulations occurred, the planning regulations have evolved into the political football that they are today.

New planning regulations completed and ready for release during my tenure were withheld on the brink of the 1996 elections. After my departure, a Committee of Scientists (COS) was appointed to offer advice. The committee's job is now complete, and the Forest Service planning staff is formulating yet another version of the regulations. And the beat goes on. The COS members, most of whom I know well and admire much, did an excellent job of analyzing and philosophizing, which was their job. But turning that philosophy into regulations is a different task altogether. One ingredient missing in the instructions to the COS was an instruction to cost it out. Cold-blooded assessment will reveal, in many cases, that the resources to achieve the stated objectives simply will not be made available.

Will revised planning regulations ever be issued? My faith is weak—unless there is a stimulus. The COS's effort has fallen on bad times because the team leader and at least one committee member have withdrawn and then returned, or have expressed serious reservations. These circumstances have weakened the political cover for the new regulations that was hoped for in the decision to appoint the COS.

In my opinion, Congress should simply refuse to fund any additional forest planning until new regulations are issued. Enough is enough—twenty-three years is long enough—a plague on both houses.

It is essential, if planning is to mean anything at all, that the disconnect between plans and budgets be corrected. Few, if any, plans have been executed as projected. I am amazed that this noncompliance with plans has produced so little uproar and so little legal action. For example, assume that a forest planning effort produced five alternatives for consideration. The third alternative is ultimately selected. However, when funding comes down, some activities projected in the plan are fully funded and others are only fractionally funded—or perhaps not at all. This is the legitimate decision of Congress and those at higher levels in the administration that allocated the budget. So, the line officer proceeds with a year's management activity on the basis of policy direction set in the budget. Has no one noticed that the third alternative is *not* being followed, or does no one care? A new management alternative is being pursued with only some semblance to the selected legal alternative.

This new alternative was not analyzed beforehand, nor were its likely consequences revealed to the public. Worse yet, whatever management is taking place is apt to change from one year to the next, depending on the vagaries of the budget. Such a process does not enhance a smooth operation nor produce a predictable outcome—not in terms of resources produced nor in desired future ecological conditions.

This situation might be corrected by requiring several budget scenarios for each alternative. The line items in the budget are arrayed at, say, three levels. Results are projected at each level. When the budget arrives, it is compared to the budget/results matrix. The level of activity related to ground or vegetation-disturbing activity is projected on the basis of the level at which any line item is the lowest. This is the application of the biological principle of Liebig's Law of the Minimum—that is, the action is limited by the weakest link in the plan (budget).

It also may be time to present the budget to Congress on a forest-by-forest basis. This plan could have one or all of several effects. First, Congress would be responsible for the distribution of funds and would get both credit and blame for budget outcomes. Second, members of Congress who had national forests in their congressional districts or states would, of political necessity, become quite interested in those entities. Third, members of Congress would become at least somewhat responsible for those national forests, thereby sharing the credit—or blame—for such activities.

Planning in its present form is quite expensive and, so far, has not yielded plans that are routinely followed. The process needs to be dramatically streamlined, and the plans that result should be followed until revised—and revisions should be made more frequently than in the past. Considering the recommendation of the COS, I have every confidence that the planning process, if ever changed, has a distinct probability of becoming even more unwieldy and costly. Coupled with the likely continuation of the disconnect between plans and budgets, it forces a serious question: Is this kind of planning worth the cost?

The Budget—Advice and Consent

It is critical that the Forest Service and its leadership be perceived as professional (that is, nonpartisan) and even-handed in dealing with Congress. This issue is particularly important when one party holds the presidency and the other party holds Congress.

The chief and the Forest Service work for and operate under the authority of the executive branch. Historically, this organization has not kept the chief and the Forest Service, a body of professionals who could be counted on for politically unbiased analysis and advice, from communicating freely with Congress. For example, until recently Congress could expect the chief, under questioning, to state clearly what the Forest Service had requested in terms of the budget, and why. Clearly, the opinion was of the chief's office, unaltered by politically appointed officials above the chief's level. Congress and all interested parties should be able to discern the chief's advice to the administration. Congress should be equally privy to such advice. The chief is and should be expected to support the budget put forward by the administration. However, I believe it appropriate for the chief to inform Congress, on request, of his or her carefully considered position on the budget. To do otherwise is to deprive Congress of information critical to its deliberations over budget matters.

I have come to consider the budget as the most significant policy document guiding Forest Service actions. Clearly, then, it is the prerogative of the executive branch to set forth its desired policy through the budget sent forward to Congress.

Understanding how the budget request is formulated is important to our discussion. First, the Forest Service assembles a budget request. That request is reviewed by the undersecretary of agriculture (and undoubtedly others), who imposes policy and emphasis to be exercised through the budget. The budget is then sent forward to the Office of Management and Budget (OMB), where an examiner (who may know a little or a lot about Forest Service matters) makes alterations and passes it back to the undersecretary for negotiation with OMB to a final conclusion. In the process, the recommendations of the Forest Service professionals have been altered, perhaps dramatically, by the undersecretary and the examiner. The logic behind the process is that it ensures the budget both reflects administration policy through line item allocations and is within designated limits.

So far, so good. But when the congressional budget committees consider the budget request, they are deprived of the advice of the chief's office. I was clearly discouraged from presenting views contrary to the administration's budget position. This development, in my opinion, is not desirable. Congress should be able to rely on the chief to give candid, forthright, and complete answers to questions from any member of Congress. Congress is entitled to the advice, assessment, and recommendations of the chief who represents the professionals that make up the agency. The administration can justify its actions related to the budget. Then, the congressional committees can do their job with full information at their disposal.

The result would be better budgets aimed to achieve long-term goals and objectives. The best means of securing such a result would be to ensure that the chief can speak freely, completely, and promptly when asked for information, opinions, and advice. Any constraint on the chief in this regard deprives Congress (whose members also were elected by the people) of information and advice critical to its deliberations.

Constituencies and Accountability

The Forest Service's organization was built, quite successfully for many decades, on an appeal to various constituencies. Those associated with the timber industry supported the timber and roads programs. Old-line conservation organizations—those associated with hunting and fishing, and some environmental groups—supported the fish and wildlife program. Livestock permittees supported the range program. Recreationists supported the recreation and wilderness programs. Mining interests supported the minerals program.

To sustain that support and play to the propensity (perhaps need) of Congress to micromanage budgets, the appropriations for the Forest Service were divided into numerous line items. Each Forest Service staff group (and corresponding staff at regional, forest, and district levels) and its supporters then had its budget, which had been networked with support groups and congressional budget committees to achieve. That budget became and remains a measure of power and influence of the disparate groups.

It has become increasingly obvious in recent years that functionalism is producing internal divisions in the Forest Service while the need for fully integrated management is being more and more recognized. Line managers have plans to be implemented. Many feel unnecessarily constrained by having to operate with more than seventy separate line items or bank accounts.

When efforts were made to provide line officers more management flexibility by creating a line item for ecosystem management that pooled money from several previously existing budget line items, there was resistance—both overt and covert—from functional staffs and their supporters. Some staff leaders actively worked behind the scenes with their constituencies to overturn that decision. Each specialty group in the Forest Service felt that it was losing power and that some other specialty group was gaining at its expense.

As it becomes more and more obvious that the move toward ecosystem management, holistic management, or restoration ecology is both real and irreversible, it becomes more obvious that the budget line items that divide the Forest Service personnel into functional groups is operationally outmoded—through still quite viable politically. In fact, it divides personnel into groups, each with its own agenda, and works against the spirit required to achieve the vision of integration of disciplines into effective teams.

Line officers faced with the responsibility of achieving the goals and objectives of operative land-use plans are almost trapped into "innovative bookkeeping" as they balance their operational checkbooks. More faith and confidence—and responsibility and accountability—should reside with line officers as they strive to get their jobs done. However, that faith and confidence should be accompanied by much-improved accountability. That accountability should be reinforced with an increase in scheduled reviews and unannounced spot checks. Supervisors who run a tight ship would, if past experience is any indication, welcome such reviews as a chance to demonstrate what their teams can accomplish and have accomplished.

In retrospect, the Forest Service once had a rigorous review process that was taken most seriously. It was a good system that should—appropriately modified—be reinstated.

In the sense of both internal and external politics, this erasure or softening of funding lines between disciplinary (functional) groups will be difficult or perhaps impossible to achieve. Distrust between the specialty groups, each with its political support groups, is significant. Some in each group routinely engage in efforts to subvert direction, decisions, or orders with which they disagree. This derision takes the form of stirring support groups to action in opposition to actions they deem inappropriate. The most pervasive action is the art of the leak, wherein constituency groups, allies in Congress, and the conflict industry receive copies of supposedly internal correspondence or e-mail. This kind of action is a violation of the canon of ethics of professional organizations such as the Society of American Foresters and The Wildlife Society, but maybe old-fashioned ethics have been replaced by "situational ethics" or "ethical adhocracy."

Functional subgroups and their support groups also fear that they will lose power and a sense of independence. This fear is most pronounced in functional groups that had to fight their way into their present positions of influence in an agency long dominated by the timber and grazing programs. These groups, such as those concerned with fish and wildlife matters, have not yet come to grips with the fact they now wield considerable influence and power. Some cling to old ways in new times that demand new ways.

More power and prestige are to be gained in helping the Forest Service establish leadership through example in the struggle to retain biodiversity and to produce goods and services in a compatible and sustainable manner. We humans must exploit our environment to survive. That issue is not in question; the question is, "How do we do it?"

Unless the folks who comprise these specialty groups can overcome their distrust and subordinate individual power to the achievement of a common and potentially much greater cause, the Forest Service will have failed its destiny. At worst, the agency will fade into history as a noble experiment that flourished for a time, then failed to evolve to fit a changing environment. It will be just another federal bureaucracy—possibly, one no longer in existence.

Much, perhaps most, of the future of the Forest Service lies with its people. If every single reform I mention here were to be magically accomplished and the distrust and maneuvering between specialty groups continued in current fashion, then self-imposed deterioration would seem possible.

Sinking into the Swamp of Litigation

One of the outgrowths of the crazy quilt of law, regulations issued pursuant to law, and case law coupled with the consequences of the Equal Access to Justice Act is a constant tattoo of legal actions aimed at Forest Service actions. Paying litigants to sue certainly encourages legal action.

Win or lose, these legal actions impose significant costs in time and money on the Forest Service (that is, the taxpayers). Conversely, when the litigants win, the issue is clarified and can guide future activities. In such cases, the litigants have done themselves, society, and the Forest Service a favor, and the litigants should be fully compensated for the costs of preparing and trying the case. The same principle should apply when the Forest Service, faced with likely loss, settles a case with a litigant. However, when the litigants lose, they are sometimes compensated. And most commonly, when the litigants lose, they do not pay the Forest Service for costs incurred in defense.

The Forest Service has learned from past errors in judgment and now is winning a higher and higher percentage of lawsuits. However, given the circumstances of low-risk lawsuits, the rate of lawsuits has not diminished. Why? First, the risk to those who litigate is low, and the chances of a payoff are great. Second, a litigation strategy that diverts agency resources away from other uses has significant "harassment value" and, ordinarily, significantly delays the proposed management action. Third, there is always the chance that, considering the monetary and time costs, the agency will both negotiate a solution and pay the litigant's costs.

Appeals and lawsuits are pursued in high numbers and now are considered a routine cost of doing business. In fact, for many, they *are* their business. This mechanism is simply too slow, too haphazard, and too expensive to be a satisfactory way to solve disputes. I suggest that such disputes be taken to mandatory arbitration before going to court. And when cases do go to court, the loser should pay the costs of the winner. It seems likely that one or both actions would dramatically reduce litigation.

Regulatory Agencies—What's Good for the Goose Is Good for the Gander

When a species or even a subspecies of a plant or vertebrate is determined by a regulatory agency (the U.S. Fish and Wildlife Service or the National Marine Fisheries Service) to be threatened or endangered, and that species is found in a national forest, then that national forest attains a co-manager. This co-manager arrangement is strange. One manager (the Forest Service) proposes, and the regulatory agency or agencies dispose through approval or disapproval. In other words, the regulatory agency can trump the land management agency in the decision.

But the playing field is not level. First, the national forest has a multiple-use mission, whereas the ESA states a single purpose: to "preserve the ecosystems upon which threatened and endangered species depend." Second, national forest managers, given the multiple-use mandate, understandably and routinely will opt for a greater risk over a shorter period. Regulatory agencies, given their preservation mandate, understandably will opt for lesser risk over a longer time frame. The result in terms of management flexibility (that is, decision space) is enormous. Third, when push comes to shove, the management agency proposes and the regulatory agency or agencies dispose.

When a management action is proposed, the burden of proof of compliance with demands (including recovery plans) is, quite appropriately, on the management agency. There is no reciprocal requirement—quite inappropriately, to my mind—for the regulatory agency to support a jeopardy opinion on the proposed management action. Because this power can and does place land management policy and action in the hands of regulatory agencies, the decisions of the regulatory agency should be subject to some form of peer review for appropriateness under the applicable science base. (Who watches the watchers? Who judges the judges?)

As is required of management agencies, the decision processes of the regulatory agencies should be transparent, technically justified, and subject to review. I do not think it was the intent of Congress, in passing the ESA, to provide co-manager status for national forests to regulatory agencies. Perhaps we should provide an appeal of regulatory agency decisions short of calling on the "God Squad." Because the God Squad is made up of such high-ranking officials and the process is so expensive, logically enough, it has been applied rarely. In other words, the ESA provides for an appeal that simply cannot be used in the normal course of events. Therefore, in reality, it is not an appeals process at all.

Three-person review panels should be established in each region to settle disagreements over legitimate disputes between land management and regulatory agencies. These panels, by their decisions, provide case law that would evolve a better vision and more consistency in applying the evolving partnership between management and regulatory agencies. Clearly, the land management agencies have the responsibility to achieve the objectives of the ESA. Should there not be a reciprocal responsibility for the regulatory agency or agencies to assist the management agency or agencies meet their multiple-use mission and objectives? I think so.

The influence of regulatory agencies over federal land management is apparent in the case of the Pacific Northwest Forest Plan. The Forest

Ecosystems Management Assessment Team (FEMAT) delivered ten options to President Bill Clinton for consideration. The so-called Option 9 was chosen, and a multi-agency team went to work on the environmental impact statement (EIS) to institute that option. At that point, the process escaped the science team, and Option 9 was loaded down with bells and whistles to satisfy the concerns of the regulatory agency or agencies and allay distrust within and between agencies.

The result of the EIS was delivered to Washington during my first months as chief. The original FEMAT effort projected a probable sale quantity (PSQ) of approximately 1.2 billion board feet per year after a three-year ramp up. The EIS, even after significant modifications, projected a PSQ of 1.1 billion board feet. It was clear—to me, at least—that the PSQ was much more likely to be one-half, or less, of that projection and that the additional requirements for monitoring and search and manage probably were impossible to execute for technical, monetary, and personnel reasons.

The plan placed default buffers on all streams that were projected to remain in place until an assessment was completed for the watershed in question. At that point, site-specific decisions would be made as to retention of buffers with what configuration and with what treatment of stands included in the original default buffers.

The super-safe default buffer system is still in place and timber yields are, consequently, dropping precipitously. When one looks at the dendritic patterns of buffers on the landscape, it takes little detailed assessment beyond visual inspection of a map to discern that the remaining spaces cannot be accessed for timber harvest and that the highest timber sites have slipped into preservation status. This result is not what was proposed by the FEMAT under Option 9 and promised in the Northwest Forest Plan.

How has this situation come about? Too many cooks in the kitchen, I say. Where is the peer review? Does compliance with the ESA require such drastic reductions in timber yield? What do we know now that the FEMAT did not know? How far do we go to mollify distrust?

OTHER ISSUES

Cooking the Books, or Keeping Score

Many attributes of Forest Service budgets and bookkeeping practices have outlived their usefulness. These procedures had, I assume, good rationales behind them when the agency's primary mission was a sustained yield or an even flow of timber to market.

The first of these procedures are the various trust funds that allow the Forest Service to retain a portion of timber sale receipts for reforestation, slash disposal, and other activities. The intent was to ensure that when a timber sale was made, funds would be available to reforest and manage the site. That seemed logical enough. But some critics maintain that these trust funds have become an incentive to sell timber. These funds were used to finance overhead and other organizational costs. The Forest Service was accused of making sales simply to support the agency. In short, the trust funds evolved into "distrust" funds. In addition, these funds are counted as costs of making timber sales. It is time to do away with trust funds. Receipts from the sales should be directed to the U.S. Treasury. Congress then, at its discretion, can appropriate funds for reforestation, management, and other purposes.

Similarly, 25% of gross receipts from timber sales and grazing activities go to the counties within which these activities take place. These payments are used to finance roads and schools and are considered as a cost of the management activity. They are considered contributions to the counties as a substitute for taxes. Perhaps the larger policy question is whether this mechanism of funding the education of essentially rural children is appropriate for the wealthiest nation in the world. Critics of this provision maintain that these payments are an incentive for elected county officials to support timber extraction and grazing.

However, as environmental concerns have reduced both timber extraction and cattle numbers, county officials are less and less enamored of this longstanding but now unstable, dwindling source of revenue. It is time for counties to receive yearly payments in lieu of taxes. This change would shift the tax obligation from timber extraction and grazing to payments to counties equal to but in lieu of taxes.

One result of such a shift would be that these costs would no longer be counted by critics as a cost of timber sales and livestock grazing. It would make the accounting for national forests more comparable with that for private lands.

Roads and Roadless Areas

Most appropriately, a review of roads and roadless areas is now under way before proceeding with building more roads, particularly into roadless areas. This activity has been both roundly praised and condemned. Yet this forceful move by Chief of the Forest Service Michael Dombeck was primarily in recognition of what already existed: reality. New road building had diminished to a fraction of historical levels, and entries into roadless areas were rare and declining because of actual or threatened appeals, lawsuits, and civil disobedience. Clearly, environmentalists had influence enough above the chief's level to force a withdrawal of any such prepared timber sales occurring within any roadless area, regardless of the operative forest plan.

The innovative aspect of the policy is a complete review of the existing road system to determine which roads should be maintained (or even upgraded), which ones should be closed to vehicular traffic (or even removed), and where new roads should be built. This review will make priority recommendations and estimate associated costs. It will not be done without conflict. The process likely will produce polarization, which is already beginning to blossom. Recreationists, whose activities are associated with road and trail use, are beginning to organize to resist closures and the decommissioning of roads.

I concur that it is simply time to acknowledge "the elephant in the room" that the Forest Service road system represents. The situation is a classic example of what economists call "externalities"—costs that are not accounted for in a transaction. In the case of roads, past Congresses and administrations continued to fund new construction to access timber and ignored the rapidly accumulating environmental effects, the maintenance backlogs, and the increasing need to maintain the existing roads.

Some wag once said, "When you find yourself in a hole, quit digging." In the past, no amount of pleading by the Forest Service brought forth a willingness to face the burgeoning externalities of a deteriorating, inadequately maintained road system. So, Chief Dombeck decided to quit digging until we analyze and face up to the accumulated road system on the national forests. Finally, the bill for fifty years of aggressive roading and inadequate maintenance is due and payable. Good for Chief Dombeck.

It is probably best to remove most, if not all, roadless areas from consideration for timber production under present conditions. The political realities of the moment already have placed a de facto moratorium on roading and timber harvests in such areas. However wise this de facto decision, we must adjust forest plans immediately to remove the longstanding illusion that timber will be available from such areas—at least within the life of the current plan. This action may result in a larger drop in the forecasted timber supply than would be expected at first blush, because rates of timber harvest are predicated on the timber in these areas being available for cutting.

When I was chief (from 1993 to 1997), I issued instructions that either we build roads in roadless areas in the timber base so the timber could be cut, or we remove these areas from the timber base by amending the forest plan. Associated adjustments to anticipated timber yields were to be announced simultaneously. For whatever reason, probably the proximity to a new round of forest planning, little was done to respond to that direction. This unfortunate delayed response to instructions reduced timber yields even more dramatically than expected. So, I applaud Chief Dombeck's decision to face up to reality—and the consequences of past actions—across the National Forest System (NFS).

Thousands upon thousands of miles of roads are the residual effect of initial timber harvest. These roads are but one example of externalities that were associated with initial timber extraction and were not considered costs at the time. Addressing such externalities (for example, channelization, damaged streams, and soil impacts) was either not recognized or put off for another day. That day has come.

The Interior Columbia River Basin Assessment (CRB) has clearly revealed the cumulative impacts of the externalities connected with grazing, timber extraction, and mining. It is also clear that restoration—that is, facing up to the extant situation—will be neither cheap nor easy. The reaction of elected officials from the region in question has been revealing. They have expressed outrage at declining levels of resource extraction from the national forests. Yet, simultaneously, they have expressed outrage at the projected costs of dealing with the consequences of accumulated externalities. One does not need a crystal ball to forecast the likely result of continued inaction. As I have heard some in Congress express it, "that dog won't hunt"—at least, not any more.

I have heard it said that doing the same thing over and over and expecting a different result is called insanity. Interested parties ought to ponder this saying relative to the current situation.

Authority Should Be Commensurate with Responsibility

The chief is appointed by the administration in power through the secretary of agriculture. Traditionally, the chief's appointment overlapped changes in administrations. Such was, to my mind, a good tradition. I hope that it is not dead.

The chief should be solely responsible for carrying out policy and directing activities of the Forest Service. Within legal and policy boundaries, the chief should be solely responsible for staffing, operations, and results. Micromanagement by administration political appointees is contrary to any set of management principles of which I am aware.

Additionally, the chief should be considered a primary expert (often through staff) on national resource management policy within the administration. As such, the chief (and appropriate staff) should be consulted by higher levels of government as resource management policy is formulated. Political policy decisions conducted without thorough knowledge (which the chief can help provide) of the technical, economic, legal, political, social, and historical ramifications of such policy are not apt to produce good results.

The chief (and the Forest Service in turn) should be consulted on any policy or activity that he or she is expected to execute. To do otherwise can and routinely does produce disastrous results. Examples that took place on my watch were the Salvage Rider and the proposed swap of national forest lands for the New World Mine.

Below-Cost Management Actions

One of the more popular political ploys used in the efforts to resist proposed land management activities is the issue of below-cost management activities, primarily forest stand treatments and livestock grazing. These ploys work well in influencing public opinion but grossly simplify very complex issues.

This complexity results largely from the rules on how to keep accounting records. The best example is the controversy over below-cost timber sales. Given the bookkeeping rules and the vulnerability of the proposed timber sale to legal action and political activity, it is increasingly unlikely that a timber sale will make money. Consider this scenario: First, suppose that payments of 25% of gross receipts are replaced by tax payments. This "cost" of a timber sale is eliminated. Furthermore, suppose that trust funds—such as Knutson–Vandenberg funds that set aside a percentage of gross receipts for future stewardship actions—no longer exist. Then, consider that some portion of the cost of the associated roads involved is marked off against recreation. Most of the recreational use of national forests is associated to some degree with forest roads, most of which were originally constructed to facilitate timber sales.

Finally, what if other actions, such as fuels reduction or thinning or production of a desired wildlife habitat condition, are an objective of the prescribed stand treatment—that is, the timber harvested is only one of several purposes for the management activity, and some appropriate allocation of costs to other benefiting functions is marked off accordingly? Such stand management activities are being referred to as stewardship sales in which the timber harvested pays for a *portion* of the costs, and some remaining portion of that cost is appropriately charged to the attainment of other values. For example, if significant costs are incurred to carry out stand management to reduce the danger to homes and human life in the forest–urban interface and the commercial material removed recovers only half the costs, is that a below-cost activity?

The result of these changes would be that more, if not most, such management activities would be closer to being above cost as far as the timber extraction aspects are concerned. The most significant part of the calculus, which determines whether such a management action is above or below costs, is the set of rules for making the calculations. One should remember that some stand management activities—say, precommercial and commercial thinnings—will almost certainly be below-cost activities, when in reality, they are investments in achieving a future desired condition. In other words, such actions are investments.

New accounting rules, coupled with the fact that all receipts would go to the U.S. Treasury (because trust funds would have been abolished), would change the picture—and the bottom line—substantially.

To the extent that cost–benefit assessments are germane, the evaluation should reflect reality as much as possible. In my opinion, the present procedures are wacko, subject to manipulation for political purposes, and badly need to be reconstituted.

Timber sales made entirely for purposes of providing wood to the market below costs are another matter. Such sales provide benefits to a select, geographically defined segment of the population. There may be policy reasons for such activities, including support to local communities and industries and maintaining downward pressure on prices for wood products. The extent to which such actions are deemed desirable is a question of policy. Congress and the administration should determine such policies—not the Forest Service.

Throwing Research Scientists into the Management Breech

Over the past decade, the national forests have experienced a burgeoning demand for scientists to help in assessment, planning, and management duties. This demand has increased simultaneously, and probably coincidentally, with a decline in the number of key experts available to national forest managers because of "meat ax" downsizing efforts that offered incentives for early retirement or departure. My general impression was that many of the most experienced and talented employees took advantage of those buyouts to change employers.

The result has been that the Forest Service (and other agencies, such as the U.S. Fish and Wildlife Service) fell back on scientists from their research divisions to head and staff high-profile SWAT (Solve with Available Technology) teams to address crises that erupted one after another. These efforts include the Interagency Scientific Committee to Address Management of the Spotted Owl (ISC), the Scientific Assessment Team (SAT), FEMAT, CRB, and the Alaska Plan and Assessment efforts.

These ad hoc science teams had credibility and capability and produced the desired results after they were given the independence, resources, and the necessary mandate to provide solutions to vicious management problems. Yet, confusion emerged as to what these efforts were and were not. Some people in power wanted to be assured that these efforts were good science. Yet, the efforts were not science at all.

They were assessment and planning exercises carried out by very highly qualified and skilled scientists—but they were assessment and planning nonetheless. These efforts required skilled scientists from an array of disciplines to consider and integrate the available science and then integrate with the science from other disciplines to produce coherent and defensible approaches to assessment and management.

Although it has been heartening to see how well these teams responded to their assignments and observe the quality of the products of such efforts, there is a significant down side. First, many of the Forest Service's best and most respected scientists were abruptly taken away from their ongoing research assignments. Many of these research efforts addressed critical questions, and the research suffered as a result. The time pressures and the involved social, economic, political, legal, and ecological consequences of these efforts were enormous and immediate. Some of these teams essentially worked ten- to fourteen-hour days, seven days a week, for three to six months. And some of the most skilled and flexible team members moved from one such effort to another and then to another without respite—and without complaint on their part.

Perhaps the crises demanded such inordinate efforts and sacrifice. But this approach cannot be allowed to become the norm without significant adverse impact on the Forest Service's research efforts and the consequences of combat fatigue and burnout on the scientists involved. It is critical to recognize that the demand for very high levels of technical expertise within the NFS will continue and likely increase. The use of research scientists in that role cannot continue indefinitely. Although it is improving, the NFS is not well equipped in terms of organization, appreciation of the problem, or philosophy to deal with this sea change. Appropriate reaction will include recognition that these crises are becoming routine and predictable. Therefore, these events should no longer be visualized as crises but as the norm.

The appropriate response is to better anticipate developing crises and to organize to meet the situations in a calm, controlled manner. Adhocracy in response to such matters should be replaced with a table of organization and standard operating procedures. The Forest Service and its co-managers now have enough experience and institutional memory in dealing with a series of such events to address this need effectively. The NFS must employ (or produce through additional training of current employees) an adequate cadre of scientists (probably at the Ph.D. level) of appropriate disciplines to address these needs. To ensure an appropriate workforce that reflects the required disciplines and experience, the table of organization will have to be coordinated nationally. The personnel required can be distributed across the NFS. Yet, the scientists and disciplines involved must provide a coherent array of talent. First, the NFS needs to cultivate current personnel by using the Government Employees Training Act authorities to upgrade the technical capability of its best and brightest who also are willing to obtain additional training and face the rigor of addressing crises. The researchers who have made up the crisis teams have by and large been of senior rank and, not too surprisingly, have performed accordingly. The pay grades of these lead scientists were GS levels 13 and 14 and ST levels 15 to 17.

Attaining—and maintaining—such a cadre of staff will require not only cultivating current personnel but also recruiting and paying such personnel under a "person in job" concept. In brief, such persons may hold higher grades than their supervisors but are considered for reward on the basis of achievement and skill level, not their positions on an organizational chart. This situation has long existed in the Forest Service research division. It is time for the NFS to follow suit.

Addressing this issue will, of course, produce the grade creep that causes consternation in some circles. But that grade creep, I submit, is inevitable as the demands for educational, experience, and skill levels continue to grow. The game has changed dramatically, and the workforce must be adjusted accordingly.

It is essential to clearly define the role of scientists in dealing with management questions. Drs. Tom Mills and Fred Everest of the Pacific Northwest Experiment Station and Phil Janik, who was Regional Forester in Alaska, have led the way to significant progress in this regard. Under their concept, scientists will ensure that applicable science information has been brought to bear and then that this science has been appropriately considered and used in management assessment and planning—the so-called science consistency check. The designated Forest Service line officer is charged with the final decision, which may well involve information not of a technical nature. But the science is clearly documented and accounted for in the final decision.

Again, all sides involved in national forest planning and management profess great faith in science and scientists; however, I wonder whether they really understand what science is and who scientists are. It should be recognized that many who express such faith have little knowledge of what science is and is not. This last bastion of faith in the continuing struggles over appropriate resource management must be protected and reinforced.

Esprit de Corps Is Essential and No Accident

The Forest Service has long been recognized as an elite organization with high levels of professional competence and esprit de corps unparalleled in government. This spirit has largely been a result of recruiting professionals of the highest quality and then investing in their careers. Leaders were identified and cultivated by training and service in positions of increasing responsibility. Essentially, qualified professionals from within the organization filled all leadership positions.

The down side to such an organization is that unless carefully guarded against, too much conditioning precludes new ideas and approaches. The total absence of new ideas, new approaches, new experiences, and new views that would accompany the infusion of talent from the outside should be recognized and tempered by astute recruitment from the outside. Balance is everything. This dark side has been diffused somewhat as a problem as the Forest Service workforce became more and more diversified by profession, gender, and ethnic status after the catalyst of the NFMA and the National Environmental Policy Act.

The up side—cohesiveness, loyalty, esprit de corps, and the feeling of being part of something greater than oneself—has been a significant part of what has set the Forest Service apart. It has allowed an evolutionary process to deal with changing times. But now it seems that times they are a-changin' faster than evolutionary processes can absorb. Conditioning the Forest Service organization to deal with the rapidly shifting technical, legal, and social circumstances is critical and may require departure from time-honored and tested ways of operation.

When political appointees above the level of the chief meddle with appointments in the Forest Service, the rank and file's respect and confidence is eroded and diminished, even destroyed. The chief should receive policy direction and be held accountable for achievement. For best results, the chief should pick his or her own team. When the chief is perceived to be surrounded by staff that are outsiders, the rank and file can jump to the conclusion that they, collectively, are either distrusted or considered incompetent. These perceptions, if not countered with explanations to the employees, can breed distrust. The advantages and the necessity of such appointments should be explained—and that should be possible.

When the rank and file know that political appointees are placed in the Forest Service, they cringe. Why? Because they don't know these people and have no way to judge their competence. Because they know their tenure is both doubtful and limited. Because, most of all, they know that it is likely a harbinger of the future. Is this what is to be expected when a new administration takes over? Where is the steady hand on the rudder? Is their icon of the chief still a valid investment of loyalty and faith? Esprit de corps is built and maintained on these intangible values: vision, loyalty, and continuity.

Is the Forest Service, then, to be just one more collection of bureaucrats carrying out politically assigned tasks? This is not the future that most Forest Service people envisioned, and they contemplate it with foreboding. They signed up to be part of the best agency in government. They wanted a career where the best of the best had the chance to rise through the ranks and/or through professional achievement to direct the outfit. The people of the United States deserve at least one agency in which those traditions of service and achievement are held high. It is a tradition that is too important to be cast aside.

With rare exceptions, the Forest Service should be staffed—particularly at the highest levels—with the best employees that the agency can produce. Stability is better established or maintained. Pride in the outfit is enhanced. Confidence in the leadership is preserved. Loyalty is instilled. And the critical esprit de corps lives and passes from generation to generation of the outfit.

To do otherwise diminishes those almost magical attributes that produced an effective organization that has sustained performance of the highest caliber over time. The management of natural resources with a vision that spans centuries requires this kind of an organization.

SOME POSITIVE SUGGESTIONS

Zoning for Timber Production?

If the national forests intend to produce a significant amount of wood for the American people, a clear direction from Congress and instructions from the administration to do so are required. The best means of producing wood on a predictable schedule and at a particular rate depends on reducing or managing the variables that impinge on management action. Such factors as markets, insect outbreaks, droughts, and fire cannot be fully mitigated. "Nondeclining even flow of timber to market" no longer seems a viable objective.

Timber should be sold when the price is right, not offered on a set schedule. No other owner of timber would market wood on a regular schedule regardless of demand and price. Forest Service timber is put on the market on a regular schedule (that is, oblivious to market demand) and allows leeway to buyers as to when the timber is cut. This practice encourages speculation among buyers, who may delay timber cutting until an economically advantageous time. And history shows that Congress has been willing to shield speculators with buybacks when things go sour. Instead, sales could be prepared and put on the shelf until an appropriate marketing time. Cutting can be required by a certain time. For the sake of workforce stability and efficiency, sales can be prepared on an even-flow basis and more opportunistically marketed.

Insect and disease outbreaks in forest stands managed for timber production can be mitigated through appropriate stand management and temporary control, by using chemical and biological control agents, until the stand composition can be altered to achieve desired conditions through fire or mechanical means. Droughts cannot be controlled, but the fire danger that increases during such periods can be anticipated and addressed through stand management, including the use of thinning and controlled fire.

Below-cost sales can be addressed by concentrating resources in areas where economical growth is most probable: high-site lands with relatively low elevation, gentle topography, second-growth managed stands, an extant road system, easy access to recreation, and minimal environmental risks. But given the time cost of money, it is a ludicrous gamble to invest the money necessary to achieve high levels of productivity of wood if there is no reasonable certainty that the trees will ever be harvested. For example, an initial investment of $500/acre at 7.2% interest would have to return $128,000 at the end of an eighty-year rotation to break even.

If the national forests are to be expected to produce timber, a "Hobson's choice" must be addressed: The best areas to economically grow timber also are apt to be the most biologically productive of both biomass and biodiversity. The alternative is to practice more extensive, or opportunistic, timber extraction from lower site lands at higher elevations that are more difficult to access; have steeper topography, higher environmental risks to disturbance, and lesser capability to produce biomass and biodiversity; and are likely to produce below-cost sales. Which way should we go? If we don't make this decision now, it makes little sense to make such investments and then accept the natural course of events.

The only way that I can see to make that choice is to zone the highest site lands with the lowest potential for environmental damage from stand treatments and roads for emphasis on timber production. Zoning has been applied to wildernesses, wild and scenic rivers, recreation areas, and so forth with some success. Why not zone timber production areas? It does not mean that multiple uses would not take place on those lands. It means that such lands would be managed primarily for the growing and harvesting of trees in a sustainable manner. These lands could be identified through planning and then established by law or some other mechanism to ensure ability to capture a return on investment.

If this change were to take place, the issue becomes whether these timber-emphasis lands should remain in public ownership. Would it be better to trade those lands to the private sector in exchange for more acreage with lower timber values but with higher values for watershed, recreation, aesthetic beauty, and fish and wildlife?

The primary argument for public ownership and agency management in today's circumstances lies in the concept and practice of multiple-use forestry. In such an approach, it is not only acceptable but expected that forests be managed in such a way that multiple uses are the focus. This practice allows—and perhaps mandates—management decisions that will not be maximally efficient in the economic sense. However, if timber and maximum economic return were the criteria for success, then it would be difficult to rationalize why the forests should remain under government management. These lands, if offered in trade on a value-for-value exchange, would likely yield a dramatic enhancement of federal land holdings for other values.

Fees for Noncommodity Uses of National Forests

It seems both reasonable and increasingly necessary for the national forests to be managed in such a fashion that revenues come as close as possible to covering costs of management. As revenues and associated economic activities related to timber, grazing, and mining decline, the gap between revenues and expenditures will expand, and the reduction in revenues to county governments will become more pronounced. This prediction assumes that revenue sharing with counties is substituted with payments in lieu of taxes.

Why should we be so concerned about below-cost timber management programs while we ignore a similar circumstance related to recreation? If current trends continue—and it seems likely—to deemphasize timber, grazing, and mineral extraction while increasing the emphasis on recreation, water, and fish and wildlife, then who pays the tab? Should those who benefit disproportionately—that is, the users—pay some significant portion of associated costs (which, of course, include both direct and opportunity costs)? I think so. But several issues must be addressed.

What about, "I already pay taxes; aren't my uses of the national forests already covered?" My answer is that those who benefit more should pay more. An analogy exists in the public university system wherein the taxpayers subsidize the system, but students who use the system pay a fraction of that cost.

The difficulty in collecting such fees is presented as a huge barrier to such an approach. Entrance fees to use national forests, analogous to those charged for national parks, are of limited use because of differences in ease of access. Why not charge a federal land use fee good for entry to federal lands—*all* federal lands—and then charge additional fees for special uses such as hunting, fishing, campgrounds, and so forth?

In addition, any opportunity costs associated with maintaining pristine watersheds that provide high-quality water for municipal uses have not been appropriately considered. For example, the city of Portland, Oregon, gets water from the Bull Run watershed—which is maintained in pristine condition—that is of such high quality that no water treatment (that is, filtration) is required. This watershed is made up of some of the most potentially productive timberlands in the world. The economic returns from timber management are forgone in the interest of sustaining a high-quality water supply for Portland. Likewise, recreational uses—including hunting, fishing, boating, hiking, and camping—are precluded. Given the proximity to the Portland metropolitan area, this decision represents a dramatic opportunity cost.

Why should the people of Portland, who are the sole beneficiaries of maintenance of the status quo, not pay these opportunity costs? Perhaps they should pay at least the initial, maintenance, and operational costs of the water treatment facilities that would be required in the case of management and use of the watershed? At the very least, the national forest that contains the Bull Run watershed should be credited with such revenues. Why should the taxpayers at large subsidize the citizens of a single city?

Of late, great emphasis has been placed on the value of recreation over timber harvest. A recent Forest Service assessment reported that recreational uses (including hunting and fishing) of the national forests produce thirty-two times the revenue of the timber program. Two points are involved with the use of this startling statistic. The first is the assumption that the timber program is antithetical to recreational use—that the two outputs are incompatible. Is that so? If so, to what extent? Clearly, the significant use of the national forests for recreation is linked inextricably to the road systems that resulted largely from the timber program. The second point is that this huge economic activity returns an insignificant amount of money to the land on which the recreational activity occurs.

The associated below-cost concerns with other forest uses beyond timber, grazing, and mining are significant and should be clearly addressed—the sooner the better. Clearly, I play the role of devil's advocate in this case. My point is that we need to rethink the questions of who benefits, who loses, and who pays for what. Should the user pay at least the additional costs associated with the use in question? I think the answer is, inevitably, yes.

Should the user be concerned as this trend evolves? The answer is, again, in the affirmative. I believe in the golden rule—that is, "He who has the gold rules." In other words, those that pay even a portion of the costs associated with their particular use of the public's lands have increased influence and increased political interests. The people who pay become true stakeholders, deserving of a place at the management table. This evolution may well be the genesis of the constituencies that supported the direction taken by past Forest Service management.

Consolidation

The possibility of reducing the number of Forest Service administrative units should be considered for the sake of efficiency and in the interest of directing more resources to the ground. I do not believe that this task can reasonably be accomplished by removing a layer of administration (Washington office, regions, national forests, or ranger districts). However, it is quite reasonable to combine some regions, forests, and districts. My personal preference would be to combine regions and forests and leave the customer service centers (that is, the ranger districts) largely in place.

Under present circumstances, combining any significant numbers of administrative units is highly unlikely, no matter how well justified in terms of efficiency. Congress simply will not allow many such actions. Micromanagement by Congress in this regard is so pronounced that permission of six—I repeat, six—congressional committees is required before the Forest Service can close a single ranger district office. So, in reality, if a single member of Congress objects to the action, the Forest Service will not receive approval. Objections and failures to carry out proposed closures are the rule rather than the exception. So why even try? The cost—monetary and political—is simply too high. No other agency that I know of is under such a constraint.

This relationship results because the Forest Service offices and their personnel are significant parts of the economy and social structure in hundreds of small towns. And Forest Service employees ensconced at local levels are adept at heading off any such actions that would cause them personal inconvenience or damage.

Under present circumstances, it is highly unlikely that any significant action can be taken in this regard. Two solutions seem possible. The first is to take Congress out of the issue via the initial suggestion of the agency being managed under a board. The second, which might gain support, is to follow the example of the Department of Defense and have Congress name a base closure board to recommend a well-thought-out package of closures—that is, combined offices—to be accepted or rejected en masse. It would be the height of hypocrisy for Congress to jab the Forest Service on matters of efficiency and then steadfastly prevent actions that would improve the situation in order to play to a local constituency. Too many members of Congress are all for government efficiency, as long as the efficiencies are recognized in someone else's district or state.

CONCLUSIONS

Shifts in Emphasis

As the emphasis for the management of the national forests shifts from traditional areas of commodity production (timber, grazing, and mining) to the preservation of biodiversity, enhancement for fish and wildlife, recreation, and watershed protection and enhancement, the budget will likely decrease. This result seems probable because it will be perceived (perhaps for reasons of political leverage) that it should cost less to handle a program of management that includes reduced commodity programs.

The Forest Service already has been threatened by powerful committee chairs in Congress that a reduction in timber production will lead to a budget predicated on custodial management. Dealing with the accumulated and previously ignored externalities of unmaintained road systems, the restoration of ecosystems damaged by past management actions may provide more economic opportunity and stimulus than building false hopes for a return to the good old days of a twelve billion board feet/ year timber program. Ecosystem restoration efforts will place emphasis on riparian zones and repair existing recreational infrastructure, create additional recreational facilities, bring trail systems up to standard, and regulate human use.

As commodities production continues to decrease and as population and resource consumption continue to increase, it is clear that the difference in timber and livestock production will be made up by state and private lands in the United States and abroad. If it proves to be so, then we should increase emphasis on efforts by Forest Service units in the arena of state and private forestry, research, and international forestry programs and offer proportionately less on management of national forests for timber and grazing. Such shifts are not only pragmatic in that pressure on those private lands will be increased to produce more commodities from an ever-decreasing land base. Given the ramifications of such a shift in forest policy, it is an equally necessary policy to provide the science, extension of information, and incentive programs to encourage both industrial and nonindustrial forest land holders to provide an increased share of the timber and livestock production in the United States in a sustainable, acceptable fashion.

It clearly follows that existing and developing knowledge and technology be extended to private land holders on a stepped-up and continuing basis. The Forest Service's State and Private Forestry Division is well-experienced and well-positioned to accelerate efforts in that regard. The Forest Service's close and long relationships with state foresters provide an opportunity to maximize the probability that private lands can take up some of the slack and practice sustainable forestry.

After a good start, the Forest Service's International Forestry Programs were slashed—in my opinion, unconscionably—by Congress in the mid-1990s. I submit that if we decide, or are compelled by circumstances, to significantly reduce production of forest products in the United States and obtain our wood more and more from imports, then we are pragmatically and morally obligated to provide technical assistance and training for the natural resource management professionals in other countries. We should assume some significant responsibility for the environmental consequences of our de facto decision to obtain our wood elsewhere. How can we, in good conscience, escape that responsibility? Too often, "elsewhere" will be relatively poor nations that have few regulations to protect the environment and ensure sustainable forestry practices, and their desperate need for American dollars will far outweigh any environmental concerns. I believe that the United States has moral, ethical, and pragmatic reasons to provide aid to those nations. The Forest Service is well-staged to organize and direct that effort.

Research is key to achieving a vision of sustainable forestry and ecosystem management. No other nation has the research capability in the arena of natural resources that the United States does, when the talents and resources of the Forest Service, universities, and private industry are combined. All the research that has gone before is but a good start on the foundation on which sustainable natural resources management will be constructed and reconstructed.

We truly are in a race between achieving sustainable management or renewable natural resources and experiencing disaster. Constant improvement in both knowledge and the application of that knowledge and evolving understanding is essential. Fortunately, the Forest Service research arm is, in my opinion, the finest research organization in the world dealing with issues of natural resources. In cooperation with universities (where strong partnerships already exist), stepped-up research can provide sorely needed knowledge to enhance the probabilities of sustainable forestry on federal, state, and private lands.

The biggest challenge that lies ahead is the cultivation of our individual and collective abilities to not only produce new insights and understanding but also synthesize information from myriad fields of interest into useful and applicable forms. If additional funds become available from any source, including reductions in funding for the NFS, then research—particularly cooperative research efforts—should be enhanced.

Who Should Manage Public Lands?

The validity of returning the federal lands to the states is under discussion. First, we should be clear that the states never owned those lands in the western part of the United States; the territories surrendered those lands to the federal government on achieving statehood. The Forest Service lands in the east were purchased by the federal government.

The concept in such a suggestion is that the states are more efficient land managers and that the federal lands in state ownership would be more efficiently managed. This comparison is predicated on a false premise: that the objectives of state and federal management are the same. But two totally different management objectives are in place. The state lands are, by and large, managed under trust responsibility, the objective of which is the maximization of revenue to the state. The federal lands are managed under the mandate of multiple use, with a new overriding mandate for preservation of biodiversity. To compare management for efficiency is to compare apples with oranges.

The appropriate test is to direct the national forests to be managed under applicable state laws. I doubt that the results in terms of efficiency would be much different, but a different signature would appear on the ground. Is that a signature that the people of the United States would agree with? I doubt it.

I once engaged a state forester colleague in conversation regarding the state's assuming management of the national forests. He allowed that he would like to give the idea a try under state law. When asked how he thought he might operate under extant federal laws and regulations, he just laughed and walked away. His parting statement was, "No way."

What would have happened to the present national forests if they had been placed in state custody from the beginning? Would much, or any, of these lands remain in public ownership? How would they have been treated over the last century? I do not like the picture that comes to mind—how about you?

I don't believe for a moment that the people of the United States will tolerate a diminution of their public land heritage. In fact, I believe that they will insist on a select increase in those holdings.

As long as those lands remain in public ownership we can and should debate their management. The question, then, is the appropriate course of action for the Forest Service, which I have already described as the best conservation organization in the world. Given a clear mandate and support, the Forest Service is the best way to get the job done.

Discussion

What Now?

Hanna J. Cortner

> The predictable conditions of the transitional decade of the 1990s, including the unpredictability of events, will require more fundamental change efforts involving large organizations than at any time in the past. We live at a time in history when the basic institutions of society and the relationships between them are being reevaluated and redesigned. The role of wealth production in the society and the allocation of wealth between rich and poor, the First and Third Worlds, and north and south are taking new forms. The relationships between governments, the so-called nongovernmental sector composed of groups of volunteers and citizens, and the producers of goods and services—the private sector—are in constant change. Anyone in a leadership position in a large organization has to be acutely aware of these changes and the challenges that result from them.
>
> —Beckhard and Pritchard (1992, 93)

The opening quote is drawn from a book about organizational change in the private sector. It illustrates that the forces prompting radical rethinking of organizational purpose, priorities, and structure; demanding new visions for the future; and altering leadership responsibilities in the private sector are strikingly similar to those that affect the Forest Service. No longer can large organizations, private or public, assume that they control their own destinies or that they operate in a relatively stable, predictable environment. Changing the essence of large organizations

HANNA J. CORTNER is professor of renewable natural resources, School of Renewable Resources, University of Arizona.

requires operating in a learning mode in which both learning and doing are equally valid. The first requirement for so doing is the "absolute essentiality of a fundamental change effort being vision-driven." Several elements must focus the change effort: changes in the mission or "reason to be," identity or outside image, relationships to key stakeholders, the way of work, and the culture (Beckhard and Pritchard 1992, 35 and 37).

As we consider the Forest Service in the context of fundamental change efforts, it is important to solicit the perspectives of former leaders of the agency regarding the forces that are pushing for change, the issues that need consideration in choosing a change strategy, and the leadership implications of moving the organization to a learning mode. Jack Ward Thomas recently was in the hot seat as chief of the Forest Service. In Chapter 2, he describes the changed circumstances under which the agency operates and presents many specific suggestions for dealing with those problems, from convening a new Public Land Law Review Commission to rethinking how science and agency scientists relate to National Forest System management. Certainly, solving many of the problems that Jack cites (for example, the veto power of the Fish and Wildlife Service over proposed Forest Service actions, micromanagement by Congress and administration appointees, the swamp of litigation, functionally based line-item budgeting, and below-cost management) depends on first defining a mission or a reason to be. Jack maintains that the agency does have a mission: to preserve biodiversity. The problem is that this mission is not stated in law; rather, it has evolved largely in response to the Endangered Species Act and the agency's own planning regulations. Whether or not he believes biodiversity should be the primary mission for the future is not clear, but he insists that the mission be clarified.

Elsewhere, Jack has written eloquently of the need to keep the national forests in public ownership (Thomas 1997). However, what he feels distinguishes the Forest Service from other federal land management agencies and private suppliers of natural resource goods and services is not clear in Chapter 2. Unlike a private corporation, the Forest Service and its leadership alone cannot define its mission or reason to be. In addition to the agency and Congress, the American people will need to be actively engaged in conversation about the vision for the national forests.

The conversation should not be relegated to a committee of experts or the professional forest policy community alone. This community has been too insular for too long. One of its principal downfalls has been its tendency to define fundamental political problems as technical problems best addressed by professional experts. We seem amazed when ordinary political processes get in the way and when political rationality is not congruent with technical rationality.

In moving toward fundamental organizational change, some necessary elements are more or less within control of the agency itself; these include changes in culture, the way of work, and relationships with key stakeholders. Agency leadership can work to ensure that the agency culture fosters a spirit of cooperation and a willingness to share power with other agencies, nongovernmental organizations, and private citizens. Once a clear, publicly sanctioned vision is in place, the agency can address gaps between the current situation and the vision, then identify pockets of resistance to change. It can organize itself so that the actions of people in the organization are aligned with the vision. It can change how the agency relates to key stakeholders. Incentives and rewards systems can be adjusted to ensure that they encourage and reward behaviors consistent with agency's vision. There must be not only vision but also commitment to the vision.

The challenges that the Forest Service currently faces also stem from society's reevaluation of its basic institutions and the relationships between them. Although some of us wish to see the Forest Service more firmly embrace ecological approaches to management that emphasize the goal of long-term ecological sustainability, we must realize that moving toward such a goal will require fundamental change in many political institutions, not only within the Forest Service (Cortner and Moote 1999). Accomplishing this task will require examining how power and authority are distributed, how social institutions shape the character of the citizenry, and how collective responsibilities are balanced with individual rights and interests. It will require building social capital through more publicly open and collaborative decision processes. It will require reexamining a range of environmental laws and policies, and—as Jack and others have indicated—the approach to this task should not be piecemeal (Thomas 1995; Cawley and Freemuth 1997). Private landowners and the producers of goods and services in the private sector can do much more than they have done to address their stewardship responsibilities. We must better link the economic and political marketplaces, rethink many traditional economic conventions and assumptions, and consider how best to mix the use of market and regulatory policy tools.

The roles of wealth production and wealth allocation (between rich and poor, the first and third worlds, and north and south) are indeed taking new forms. We are increasingly tied to a global economy and a global politic. How we approach the question of income inequalities, at home and abroad, will affect why we manage forest resources and for whom. Events in distant lands affect our nation's ability to both preserve biodiversity and satisfy our voracious consumer appetites. Similarly, as Jack points out, the choices that we make regarding our nation's national forests have moral consequences for people and resources in other nations.

The Forest Service has to be one of the most studied federal agencies in the scholarly literature. It has spawned classic studies in public administration and has been the subject of comparative studies with other natural resource organizations and policies. The attitudes of agency employees in a time of change have been the subject of countless studies. The Forest Service has undergone major agency and congressional assessments of its performance of some of its major statutory obligations. A legacy of writings and remarks by former chiefs reflect on the problems and opportunities they encountered during their tenures, and others have written a wealth of information about the history of the agency. Numerous policy suggestions have been made for effecting organizational change in the field of natural resource management in general and in the Forest Service in particular. These suggestions involve fundamentally different political choices that define the relationships among humans, ecosystems, science, and democratic governance.

Unfortunately, the resilience and integrity of many of the political institutions through which we must act to make collective decisions are increasingly in question. Public trust of government has declined markedly over the past twenty-five years. Although several innovative community-based collaborations have emerged, forms of political participation such as voting are down, and there is serious debate about whether an overall decrease in civic participation in nongovernmental associations is eroding social capital. Traditional mediating structures such as political parties have been weakened. The strong presidency is gone and the institution damaged; Congress is polarized and characterized by gridlock. Have key democratic governance structures become so debilitated that they lack the capacity for timely innovation?

We don't lack ideas about what needs to be done with the Forest Service; we lack the political will to deal interactively with questions of democratic governance and a vision-driven change strategy.

REFERENCES

Beckhard, Richard, and Wendy Pritchard. 1992. *Changing the Essence: The Art of Creating and Leading Fundamental Change in Organizations.* San Francisco: Jossey-Bass.

Cawley, R. McGreggor, and John Freemuth. 1997. A Critique of The Multiple Use Framework in Public Lands Decisionmaking. In *Western Public Lands and Environmental Politics,* edited by Charles Davis. Boulder, CO: Westview Press, 32–44.

Cortner, Hanna J., and Margaret A. Moote. 1999. *The Politics of Ecosystem Management.* Washington, DC: Island Press.

Thomas, Jack Ward. 1995. *The Instability of Stability.* Paper presented at the Landscapes and Communities in Asia and the PNW Conference, Missoula, MT.

———. 1997. Devolution of the Public's Land—Trading a Birthright for Pottage. *Renewable Resources Journal* 15(2): 6–10.

3

The Next Decade of the Forest Service

Does the Past Hold the Key to the Future?

Christopher A. Wood

Where or what will the Forest Service be in ten years? Emblazoned on the wall of the National Archives is the phrase, "What is past is prologue." So, before we look ahead, let's take a trip back in time.

In 1989—about a decade ago—the information superhighway was a one-lane road. George Bush was inaugurated as the forty-first President of the United States; two years later, he would have the highest popularity rating of any modern-day president, and in two more, he'd be out of a job. Madonna and Sean Penn were an item. Michael Jackson was "thrilling" pop music fans. The Washington Redskins football team dominated its conference. And almost twelve billion board feet of timber were harvested from national forests.

In those days, the Pacific Northwest Region of the Forest Service alone produced four to five billion board feet of timber per year. One forest, the Willamette, sold about 900 million board feet. That year, 1989, marked the end of a thirty-year period where timber harvest levels off of national forests ranged from nine to twelve billion board feet.

In the next decade, much changed. Timber sales declined by more than 70% to about 3.4 billion board feet in 1998. The Pacific Northwest Region's timber sales in 1999 approximated what the Willamette alone had sold ten years earlier.

CHRISTOPHER A. WOOD is the senior policy advisor to the chief of the Forest Service at the U.S. Department of Agriculture.

The cause-and-effect relationship—an obvious generalization—was simple: The more trees the Forest Service sold, the more revenue it generated; the more revenue it generated, the more staff the agency could hire and the more programs it could provide to meet the demands of the American people. As long as the Forest Service provided the wood fiber, the budgets, staffs, and programs would grow through congressional appropriations and trust fund receipts. Most important, this is what society wanted from its forests.

WHERE PUBLIC AND RESOURCE VALUES MEET

Having briefly reviewed the past, let's turn to the issues that help shape public values today. Most people do not view public forests and grasslands solely as warehouses of commodities to be brought to market. Instead, they assign greater value to the positive outcomes of forest management, which take shape as ecologically sustainable goods and services, cleaner water, better habitats for fish and wildlife, and stable and productive soils. In fact, results of a recent Roper poll show that about 70% of people surveyed say that when a compromise cannot be achieved, environmental protection should be a higher priority than economic development.

The American people not only talk about supporting conservation and wise management of natural resources, they are using the ballot box to prove it. In the November 1998 midterm elections, some 125 states and municipalities approved ballot initiatives to protect open space and conserve natural resources. Voters asked—they actually asked—lawmakers to spend an additional $5 billion to $8 billion of their money on conservation measures.

Planning for the future requires a fuller understanding of how the Forest Service influences and is influenced by the public land values that people care most about. Consider the cases for recreation, water, and habitat.

Recreation

More people recreate in the National Forest System than on all other public lands. Outdoor recreation allows an increasingly urbanized society to reconnect with the land and provides real economic opportunities. Recreation in national forests could contribute more than $100 billion to the gross domestic product by 2045.

Water

About 25% of the drinking water that comes from surface sources flows across national forests and grasslands. The U.S. Environmental Protection

Agency (EPA) estimates that national water filtration and development needs could cost local communities more than $140 billion over the next decade. Imagine how much money these communities could save in water costs, not to mention flood-control costs, if watersheds performed their most basic functions of catching, storing, and releasing water.

Habitat

Of the 327 watersheds in the United States that The Nature Conservancy identified as critical to the conservation of aquatic biological diversity, many more than half are influenced by or contained within national forests and grasslands. The National Forest System lands and waters are uniquely positioned to anchor the recovery of species, providing greater flexibility for management of adjoining state and privately owned lands.

Some analysts argue that changing public values and management direction have caused the Forest Service to lose sight of its mission. Yet, efforts to protect and restore watersheds are validated—in fact, they are mandated—by the agency's century-old organic legislation.

Changes during the past decade may reflect a stricter allegiance to the Forest Service's mission, which is "caring for the land and serving people." We, the stewards of the national forests, better care for the land when we use the best available science and information to conduct management activities. And we better serve people when our management accurately reflects public values.

A NEW AGENDA

How does past experience influence the agency today? Well, for one thing, we know that yesterday's incentive system doesn't work so well today. Its legacy, however, remains pervasive. For example, most of the debate still revolves around outputs from the National Forest System (for example, how many board feet or recreation visitor days the agency produced) and inputs to the National Forest System (that is, how my favorite programs are funded relative to my least favorite).

This legacy makes for lively debate and exciting times come appropriations season, but its value in promoting conservation stewardship is questionable. Aldo Leopold's essay, "An Ecological Conscience" (1987), comes to mind:

> Everyone ought to be dissatisfied with the slow spread of conservation to the land. Our "progress" still consists largely of letterhead pieties and convention oratory. The only progress that

> counts is that on the actual landscape of the back forty, and here we are still slipping two strides backward for each forward stride.

Forest Service policies ought to promote incentives to maintain and restore Leopold's landscape of the "back forty." It helps to explain why the Forest Service took on the issue of roads and roadless areas. How could the agency ask managers to err on the side of land health while allowing the construction of new roads into roadless areas, knowing that it cannot afford to maintain the existing road infrastructure?

The road system in the national forests is more than 373,000 miles long, long enough to go around the world fifteen times—or is it to the Moon and back fifteen times? Either way, it's big. Too big. Let me explain what I mean. First, the Forest Service has a $8.5-billion backlog in road maintenance and reconstruction. Second, the annual agency road maintenance needs cost approximately $500–600 million. And in fiscal year 1999, do you know how much the Forest Service received for road maintenance? Only $99 million.

Roads are only one part of the Forest Service's Natural Resource Agenda. The others are Watershed Health and Restoration, Sustainable Forest and Grassland Ecosystem Management, and Recreation. The foundation for this agenda was laid by previous leaders such as Dale Robertson, who introduced the term "ecosystem management" to the agency, and Jack Thomas, who saw its principles implemented in the Pacific Northwest and elsewhere.

The Forest Service agenda translates to policy that affects "the landscape of the back forty" in numerous ways.

- Historically, agency managers often were rewarded based on production of outputs—timber, forage, recreation, and so on. Today, the Forest Service is instituting measures of land health that can be determined in the field and used to reward employees, develop budgets, and base planning and policy priorities.
- Forest management emphasis is shifting from what is taken from the land to what is left behind. As this occurs, the Forest Service is learning that the traditional timber sale contract can be an unwieldy tool to use in accomplishing stewardship goals. New tools such as stewardship contracting may help us accomplish management objectives.
- The Forest Service proposes that the U.S. Congress sever the connection between 25% fund payments and timber harvest on federal forests. Historically, 25% of timber receipts has gone to local counties for school and road funding. As timber harvest has declined, so have these payments. Why should the wealthiest nation in the world hold the education of its rural schoolchildren hostage to this method of

public forest management? The administration's proposal would stabilize county payments through a permanent appropriation, allowing local officials to better plan for school and road funding.

- In addressing incentives, the Forest Service cannot overlook trust funds. The agency finances a significant portion of the organization with timber receipts. Funding an organization on the back of a program whose "productivity" has declined by 70% in less than a decade is a dangerous business. To help the general public understand how investments in forest management result in tangible benefits, the Forest Service is considering opening its trust funds to greater public oversight and scrutiny.

WHERE DO WE GO FROM HERE?

I have looked to the past and talked about the present. So, what does the future hold for the Forest Service?

Not long ago, a conservationist for whom I have a great deal of respect said to me, "We'll be satisfied once we get the national forest timber harvest down to about two billion board feet per year." Had he been asked a decade ago what the harvest level should be, he'd likely have said nine billion or seven billion, or maybe even five billion board feet per year. But 3.4 billion? No way. Similarly, a decade ago, the timber industry probably could have settled for legislation that would have reduced harvest in the Pacific Northwest to two billion or three billion board feet. Both proposals were summarily rejected. Today, the Forest Service harvests about one billion board feet in the Pacific Northwest.

Looking at the past in this way speaks to the dilemma the Forest Service finds itself in today. A lot has changed in the past decade, from political fortunes to technology, tinsel-town marriages, and musical tastes—everything, it seems, except the debate over forest management, which too often is driven by outdated models from a bygone era.

Applying yesterday's arguments and debate to a changed Forest Service has insidious effects. First, it can perpetuate distrust and division, both internally and externally. Second, it can stifle dialogue and consensus. Third, it can compromise the agency's ability to move beyond sloganeering and act as a conservation leader. Because the Forest Service finds itself the target of demands from many interest groups, we may become loath to exert leadership on difficult resource management issues. Rather than face the political firestorm that often accompanies new ideas and innovative thinking, would-be Leopolds, Rachel Carsons, and Bob Marshalls (all former bureaucrats themselves) may decide to heed the advice of writer Richard Nelson (1991):

> The most I can do is strive toward a different kind of conscience, listen to an older and more tested wisdom, participate minimally in a system that debases its own sustaining environment, work toward a different future, and hope that someday all will be forgiven.

Let's hope that this scenario does not represent the future of the Forest Service. The natural resource challenges of the next decade are far too consequential for us to abrogate the agency's century-long tradition of conservation leadership.

Changes in agency management during the past decade or so demonstrate the agency's reinvigorated commitment to a strong national forest land ethic. What's missing, however, is the recognition that in the absence of a national consumption ethic, the Forest Service land ethic only shifts environmental problems to lands governed by more lenient environmental protections. For example, the demand for the 8–9 billion board feet formerly harvested from national forests did not disappear. It simply shifted to other places.

Here are some hard facts for you to consider. Between 1991 and 1996, U.S. softwood imports from Canada rose from 10.5 billion to nearly 18 billion board feet per year, increasing the pressure on the old-growth boreal ecosystems of northern Quebec. Today, the harvest of softwood timber in the southeastern United States exceeds the rate of growth for the first time in at least fifty years. The average size of homes in the United States increased from 1,520 square feet in 1971 to 2,120 square feet in 1996; meanwhile, the average family size has decreased. And the population of the United States, which represents about 5% of the world's total population, consumes well more than 25% of the world's industrial wood. If these statistics don't move you, look at the number of folks commuting to Washington, DC, in sport utility vehicles every day. Who—or what ecosystem—bears the environmental cost of all that needless energy consumption?

My point is that the Forest Service cannot talk about these and other significant natural resource issues while fighting backfires. Taking care of "other business" steals the agency's ability—and can weaken its resolve—to act as a conservation leader. For example, fifty years ago, the Forest Service was a leader in cutting-edge arguments against clearcutting and for better management of nonfederal forests. Today, the agency makes paltry investments in international forestry.

It is unlikely Americans will ever again see timber harvest levels of a decade ago taken from the public forests. Nor should they. Yet, Americans must be willing to slow consumption rates of natural resources if the national forest land ethic can extend over state lines and private boundaries, vast oceans, and to other nations of the world. These are *your* lands, and rather than pretend to have the answers, let me ask this question: What do you think the future holds for the Forest Service?

- Is the destiny of the agency determined by pitched battles over individual timber sales? Or, is it pulling together to put displaced workers from resource-dependent communities back to work repairing and restoring watersheds?
- Are the most important forest conservation debates truly about exemptions or inclusions to an interim roadless policy? Or, are they finding incentives to discourage the development of private forest land and the loss of 7,000 acres of open space per day?
- In arguing over below-cost timber sales, could a system where each forest must "pay its own way"—with all the attendant risks of privatization, dominant use, and so on—really work? Or, would it be wiser to nurture a climate in which citizens and their elected representatives understand and support the imperative of making investments in the health of national lands and waters?

In my opinion, the Forest Service of the future must reassert itself as a conservation leader into the most pressing environmental challenges of the day. Research and development should provide technologies and tools to help with such issues as slowing the international demand for wood fiber and the associated pressure placed on forests worldwide. State and private forestry should accelerate the amount of education and stewardship assistance provided to interested and willing private landowners, state managers, and urban residents. And the National Forest System should provide an international model of how to manage forests and grasslands in an ecologically sustainable manner.

The issues that face the Forest Service represent a microcosm of a more global challenge: learning to live in productive harmony with the lands and waters that sustain us. Meeting this challenge will require both personal and national investments in land health and conservation. The national track record is spotty on this count. Federal spending on natural resources and the environment as a percentage of total domestic spending is half of what it was in 1962, but I believe that taxpayers have demonstrated their willingness to invest in good stewardship if the Forest Service is held accountable for results on the land.

So, how do we get there from here? Many choices remain to be made, but two basic ideas would move the debate forward in a constructive manner. First, the Forest Service should adopt a more collaborative approach to management of natural resources so that everyone will understand the benefits of conservation and the consequences of inaction. Second, as the Forest Service refines its land ethic, the agency must work with the public to institutionalize a national consumption ethic.

The Forest Service calls this kind of approach "collaborative stewardship." To see it through to fruition, the agency must loosen up a bit on the reins. As public servants, Forest Service professionals must become more adept at talking about issues such as old growth, wilderness, naturalness, consumption rates of natural resources, and other value-laden issues that they inherently avoid. This approach requires working with a broader array of interests than the Forest Service is used to. It takes time and can be frustrating. It requires a little less righteous indignation and a little more patience and humility.

THE PAST AS PROLOGUE

What will the next ten years bring? Well, I have little doubt that the Redskins will be back on top of their conference, and I'm pretty sure that Madonna and Sean are through. However, whether the Forest Service has the wisdom to focus less on the fight and more on the progress of the back forty is an open question. And whether it has the foresight to promote a consumption ethic to accompany its land ethic remains to be seen.

Imagine what would happen if the authors of this book invested half of the energy we spent debating this topic in developing a shared vision for ecologically sustainable forests and grasslands. Couldn't we increase national investments in conservation if we lowered our voices and worked together toward common conservation goals? What if we spent the time building relationships that we spend working on a pithy putdown for the local newspaper?

One century of forest conservation has come to reflect a uniquely American sense of optimism. A confidence that the Forest Service can learn from its mistakes. A conviction that it can make short-term sacrifices to advance long-term gains. A commitment that indeed, we should leave for our children a better place than the one we inherited from our forebears. That is my vision for the Forest Service, both today and ten years from now.

REFERENCES

Leopold, Aldo. 1987. "An Ecological Conscience," in *A Sand County Almanac, and Sketches Here and There.* New York: Oxford University Press.

Nelson, Richard. 1991. *The Island Within.* New York: Vintage Books.

4

Rethinking Scientific Management

Brand-New Alternatives for a Century-Old Agency

Robert H. Nelson

Ideas are more important than many practical men and women of affairs believe. Ideas shape institutions and give them social legitimacy. The failure of an idea can mean the demise of an institution in the long run. Ideas motivate and define the culture of organizations. Without a clear idea of mission and purpose, an organization risks the decline of employee morale and commitment. For all these and other reasons, the fate of the idea of the scientific management of society, introduced into American life in the progressive era early in this century, is of great importance to the Forest Service.

Gifford Pinchot, who founded the Forest Service in 1905, was more than a technical forester. He was a leading member of the progressive movement of his time and eventually became governor of Pennsylvania. His view of the role of the Forest Service in American life—which would be adopted as the agency's core mission long after Pinchot was gone—was shaped by broader progressive attitudes and assumptions. The Forest Service was to be an instrument within the overall project of the scientific management of American society. The profession of forestry would staff the Forest Service with the requisite experts to fulfill this mission. Indeed, the creation of a forestry profession in the progressive era was

---

ROBERT H. NELSON is a professor at the School of Public Affairs, University of Maryland, and a senior fellow at the Competitive Enterprise Institute.

part of a widespread movement toward professionalism that yielded not only the Society of American Foresters in 1900 but also the American Economic Association in 1885, the American Sociological Association in 1905, the American Planning Association in 1909—new professional groups for almost every area of expertise that would be required for the future expert management of American society.

However, as it navigates the turn of another century, the Forest Service lacks a clear sense of direction and mission. The management of the national forests is in a state of paralysis and gridlock. Part of the reason is that the American public no longer accepts the main tenets of scientific management, whether applied to forests or other areas of American life (Nelson 1995b). The level of public trust in professionals, from doctors to lawyers to economists to foresters, has been declining for a quarter of a century. Before agreeing to surgery, patients solicit second and third opinions from other doctors as standard operating practice. One observer found that "judges increasingly worry that parties to a case can find an 'expert' to testify to anything and that flawed or distorted research—'junk science'—can mislead jurors" (Biskupic 1997). Other signs of this trend are seen in the willingness of members of the U.S. Congress as well as the judiciary and executive branches to intervene in the management of the national forests.

If the history of the Forest Service in the twentieth century has been closely tied to the fate of scientific management, then the future of the agency in the twenty-first century will closely reflect ideas that replace scientific management. Part of the reason the Forest Service is confused about its mission is that American society is no longer sure about the future of professional expertise and the place of science in government. The agency needs a new paradigm, but it cannot develop such a vision on its own. It confronts a range of views that extend from libertarians, who wish to abolish most of government, to active proponents of powerful new government controls as the only way to avert environmental disaster.

Thinking about the future requires a firm understanding of the past. What were the basic ideas of scientific management? Why are these ideas widely rejected now, at the dawn of the twenty-first century?

SCIENTIFIC MANAGEMENT BASICS

In a 1995 article in the leading journal of administrative science, *Public Administration Review*, Eliza Lee surveyed the interaction of political science, public administration, and the rise of the American administrative state. The leading political scientists of the early twentieth century, including Francis Lieber, John Burgess, Woodrow Wilson, Frank Goodnow, and W. W. Willoughby, articulated a "specific ideology of science and progress." It was based on the core idea that "science was regarded as the method of understanding and controlling changes." Great faith was "invested in science and technology as the engine of progress." From this beginning point, the progressive theorists "provided the ideological and institutional apparatus for the rise of the administrative state" in the United States in the twentieth century, with the Forest Service as a leading example (Lee 1995, 539, 541).

Scientific management was a theory about the capabilities of not only scientific knowledge to transform the physical world but also the political institutions by which this knowledge would be put to use. The theory of "scientific management redefines what had hitherto been political problems as management problems, the solution of which is governed by the logic of science." That is, scientific management sought "the establishment of science as the institution of governance and the centralization of power in the hands of scientists." This was possible in progressive thinking because the very processes of government administration were regarded as "objective, universal, natural, altogether devoid of historical and cultural contexts, and dictated only by scientific laws" (Lee 1995, 543).

An inevitable tension was created with the traditional precepts of American democracy; how could "government by the people" be replaced by government by a new professional elite? Progressives saw their efforts as a correction of the failures of American democracy in the late nineteenth century, when government had become the captive of big business and other special interests, if not outright corrupt. To many people, Wall Street appeared to own the U.S. Congress. In the early twentieth century, more general "disillusionment about the rational capacity of the people" abounded in matters of governance. Thus, it was imperative that objective knowledge, as discovered by physical and social scientists, be applied more directly to the processes of government. As Lee commented (1995, 541–542), progressive theorists such as Pinchot saw themselves engaged in a project in which the stakes were nothing less than "the survival of American democracy [that] rested on the use of scientific knowledge as a technology of governance."

The Forest Service was the product of the application of these progressive ideas to the problems of forest management. It should be an organization run by professionals well separated from politics. This separation would allow foresters to put science to use in the national forests in the service of "the public interest." Foresters would calculate the optimal age of tree harvest, the appropriate level of timber cut, the resources needed for suppressing forest fires, the appropriate level of livestock grazing in the forests, and many other "technical" matters. Social values, as dictated by the U.S. Congress and other parts of the political process, might enter in setting certain broad goals for the use of the national forests. But political influences must not enter into the details of forest management by which these goals were realized. The chief of the Forest Service must be a forestry professional, insulated from politics, who would supervise the overall application of scientific forestry knowledge to achieve forest outcomes in the most efficient manner possible.

From this perspective, the mission of federal foresters was to manipulate wild nature to make it "perfect." According to scientific management, Nancy Langston wrote in *Forest Dreams, Forest Nightmares* (1995), a well-managed forest should be "efficient, orderly, and useful." Federal foresters followed in the tradition of European silviculture, which "had as its ideal a waste-free, productive stand: nature perfected by human efficiency." The early foresters who set out on this path and saw the Forest Service as the exemplar of all that it represented came "with the certainty that they could use science to fix the forests, and that with the help of science, they could do no wrong." They were very "self-conscious of their mission, and proud of their new scientific discipline" (Langston 1995, 5, 8, and 10).

The practitioners of scientific management regarded knowledge obtained through formal research and other professional methods as authoritative and tended to dismiss local knowledge grounded in practical experience. The great advantage of science was that, through the results of expert investigations, it was possible to eliminate the waste and confusion that had inevitably attended to all the failed local experiments of the past. In short, the efficient method of science would supplant the haphazard, wasteful method of local trial and error.

These views are well-illustrated by the case of forest fire policy, a central issue for the Forest Service almost since its founding. In the Blue Mountains of northeast Oregon and southeast Washington State, for example, the Forest Service knew that local Native Americans earlier had frequently burned the forests there. Some of the earliest European arrivals had begun to imitate the Native American practice. But the Forest Service in the progressive era rejected this practice as simplistic and old-fashioned. Indeed, as Langston wrote, they saw "burning as part of [the old] irresponsible laissez faire logging practices—practices utterly opposed to scientific sustained-yield forestry." Moreover, "if light burning was an Indian practice, then by definition it was superstition, not science," and thus could never be an option. Instead, as Langston explained, "scientific control and management ... were the goals of government scientists." By means of science, the Forest Service would be able to "improve nature," not merely preserve it or protect its environmental quality (Langston 1995, 249, 250, and 252).

William Greeley, who later became chief of the Forest Service, said in 1911 that "firefighting is a matter of scientific management, just as much as silviculture or range improvement" (quoted in Pyne 1997, 185). This concept led federal foresters to reject even the views of many local non-Native Americans who considered the government plan for eliminating fire to be "absurd." To be sure, there were also self-interested motives at stake. Langston continued, "If light burning was accepted, that would threaten the Forest Service's very justification for managing the Forests—a justification which came from its claims to technical and scientific expertise." The Forest Service "was desperate to defend its own authority as manager of the federal forests, and fire suppression was one way to do that" (Langston 1995, 250, 251, and 253).

In his wide-ranging writings on fire policy, Stephen Pyne also illustrated how Forest Service thinking about forest fires was derived from the broader themes of the progressive era. The aspiration to scientific fire management early in this century involved "nothing peculiar ... nothing idiosyncratic to foresters." Rather, it was a natural outgrowth of the prevailing general "precepts of progressivism, the belief that scientific knowledge was essential and adequate, that public policy and public lands should be administered by experts trained in scientific management and shielded from political corruption and public whim." In the progressive gospel, the Forest Service efforts "reified" the widespread hopes to control nature for human use—"to wage a sublimated war on the forces of nature" (Pyne 1997, 185 and 192).

Applied in the specific area of forest fire policy, Pyne (1997, 187) explained,

> Against folk wisdom, [professional foresters] proposed science; against laissez-faire folk practices, they argued for systematic regulation of burning that would support, not confront, professional forestry; against the self-evident waste of fire—not only what it directly destroyed but those benevolent "forest influences" that it indirectly laid to waste—they conjured up a vision of conservation, the rational, industrially efficient exploitation of natural resources.... Fire fighting was the pragmatic merger of idealism with reality by means of applied science.

To be sure, the Forest Service in the 1990s finally was compelled to acknowledge that this application put much too great a burden on the methods and degree of knowledge attainable through the science of forestry. (For the story of a similar failure of public land management based on similar causes, see Nelson 1983 and Tarlock 1985.) Contrary to the tenets of scientific management, the most important forest fire lessons of the twentieth century were learned through practical experience and trial and error, not academic research. Suppressing fire in the past led to increased fire risks in the future, owing to the buildup of brush and dense thickets of smaller trees in many forests. As Pyne related, it took many decades—culminating in the large fires of the past decade—to prove to disbelieving Forest Service eyes that "the decision not to burn can be as ecologically fatal as promiscuous burning." In "removing anthropogenic fire from many environments," the Forest Service committed "less an act of humility than of vandalism" (Pyne 1997, 253).

Some years ago, Charles Lindblom and David Cohen (1979) made a general distinction across a wide range of professional fields between "professional social inquiry" and "ordinary knowledge." They argued that the members of professional groups exhibited certain general tendencies of thought, treating knowledge as authoritative only when it was the product of formal research, at least appearing to be grounded in the application of the scientific method. In the real world, however, they argued that professional methods typically could offer little more than "a supplement to ordinary knowledge" (Lindblom and Cohen 1979, 35).

Moreover, ordinary knowledge was not developed and brought to bear in the formal and systematical fashion expected of professional social inquiry. Rather, it accrued almost haphazardly in the course of various societal and political interactions that somehow came together to yield a decision. In an earlier article, Lindblom (1959) had famously characterized this form of government decisionmaking as "the science of muddling through." As Lindblom and Cohen then argued (1979, chapter 4), the tendency of professionals to overestimate the power of their own methods often rendered them incapable of seeing the essential contributions that other forms of knowledge and of decisionmaking could make. Although they did not have forest fire policy in mind, the actions of the professional foresters of the Forest Service over the course of the twentieth century have well illustrated this much broader failure of scientific management concepts of professionalism in the twentieth century.

A MORAL CRUSADE

In progressive thinking, the idea of progress was conceived in material terms. If progressivism was the "gospel of efficiency," as Samuel Hays (1959) described it, then the importance of efficiency lay in the fact that it put American society on the path of maximum economic growth. An efficient economy was effectively solving the material problems that had preoccupied most people for most of human history. Life had always been a struggle to find adequate food, shelter, and protection from disease. Progressives were convinced that providing for human beings' basic needs (that is, removing "struggle" from the material economy equation) would bring about a whole new era in human affairs.

As Pinchot often emphasized, the preeminent role of the national forests was to supply wood to meet the home-building needs of the nation—to solve the shelter problem at long last. In this way, the Forest Service would be doing its part in the grand progressive project of opening the way to a virtual heaven on Earth. (For a full treatment of the tenets of the progressive gospel, see Nelson 1991.) Indeed, Pinchot declared in *The Fight for Conservation* (1967) that his efforts for conservation had been designed "to help in bringing the Kingdom of God on earth."

Pinchot's thinking in this regard was very much in tune with that of other leading progressive intellectuals of the time. The social gospel movement of the late nineteenth and early twentieth centuries was characterized as seeking the "social salvation" of mankind, as having the overall objective to achieve "the coming to earth of the kingdom of heaven" (Hopkins 1940). Richard Ely, the founder of the American Economic Association, was another prominent member of the social gospel movement in the late nineteenth century. He believed that "Christianity is primarily concerned with this world, and it is the mission of Christianity to bring to pass here a kingdom of righteousness" (Ely 1889). Economics was critical to this mission because it would provide the necessary expertise to build heaven on Earth. In the narrower sphere of the management of the forests of the nation, the profession of forestry should also play its part in this grand project.

Thus, progressivism not only was concerned with material matters but also was a grand undertaking of spiritual renewal. Even though the details of how to achieve economic progress differed among the many "religions of progress"—for example, American progressivism, Marxism, European socialism, and Herbert Spencer's social Darwinism—all the progressive doctrines shared a common conviction: that the end of material scarcity would bring an end to human conflict over resources and a broad solution to longstanding problems of the human condition.

These convictions, to be sure, never were subject to any scientific analysis. American foresters never applied the scientific method to ask whether the high hopes for science and progress were objectively true. Rather, in the progressive era, many people thought it impossible to believe otherwise. The progressive vision was simply an article of faith. Scientific management was the secular religion of not only the Forest Service and professional forestry but also the American welfare and regulatory state during much of the twentieth century.

A CRISIS OF FAITH

The problem today is that most of the progressive articles of faith have not stood the test of time in the twentieth century. Enormous progress has been made in improving material conditions of life in the most developed parts of the world—for practical purposes, supplying most essential needs. Yet, the end of material scarcity has not brought the new degree of happiness and emotional well-being in human affairs that the progressives so confidently expected. Indeed, even as a nation such as Germany made great strides economically and scientifically in the first half of the century, it simultaneously plunged the world into the horrors of war and genocide. The great rise of material wealth in the United States over the course of the twentieth century has not abolished crime; the prisons of the United States today house an unprecedented 1.5 million convicted criminals.

Science has become a double-edged sword in ways that the progressives never anticipated. Scientific knowledge can greatly expand our ability to manipulate the natural world, but this power over nature can be used for ill will as easily as for good. At the horrific extreme of science's capability, the atom bomb raised the specter of mass elimination of the human race. Modern chemicals serve valuable industrial purposes but also threaten the environment. Developments in genetics such as the ability to clone animals challenge traditional ethics. These "advancements" are particularly frightening in light of the weaknesses of the political process in many nations of the world.

Outside the physical sciences, the high expectations of progressives for the development of scientific knowledge also have not been realized. All too often, social and administrative scientists have affirmed scientific truth only to discover later that they were in error. For decades, the need to control forest fires was proclaimed as a scientific truth of professional forestry, but in the 1990s, new expert thinking rejected the basic tenets of past fire policies. The even-flow interpretation of sustained yield—the "scientific" basis by which the Forest Service set timber harvest levels for several decades—today appears foolish and misconceived, as timber harvests have plunged to one-third of former levels.

Partly because science has been unable to uphold its end and partly because of a basic tension with democratic theory, the progressives' high hopes for a clear separation of politics and government administration were another victim of twentieth-century history. Today, the Forest Service is run by political appointees, rather than forestry experts. The Clinton administration has merely accelerated a longstanding trend in this regard.

INTEREST-GROUP LIBERALISM

Indeed, leading political scientists such as David Truman (1951) made their reputations in the 1950s by writing about the failure of the progressive plan for scientific management of American government. Interest groups were involved at every step of the way; perhaps there was no other possibility in a democratic system. The economist John Kenneth Galbraith (1956) wrote of the great benefits to American society from its system of "countervailing powers." The social legitimacy of government actions would no longer be determined by a claim to scientific truth. Rather, what was important was to bring the involved parties together, let them bargain among themselves, and reach a common agreement. The presence of the affected interests at the table and the ability to agree—rather than the substance of the agreement—became the decisive point.

In the late 1960s, Theodore Lowi (1969) characterized this emerging consensus within political science as "interest-group liberalism." Professional foresters were slow to appreciate the new thinking about American government, but in fact, interest-group liberalism represented a decisive rejection of the very foundation on which the Forest Service had been built and on which its social legitimacy continued to rest. For Pinchot, the "special interests" were not to be incorporated into the basic processes of governance but were forces of virtual "evil" to be excluded from objective "scientific" government.

The Forest Service partly resisted the application of interest-group liberalism in its domain but could not stand against the tide. New laws in the 1970s such as the National Forest Management Act (NFMA) of 1976 required extensive public participation in planning and other decisionmaking processes. The National Environmental Policy Act of 1970 (NEPA) prescribed a governmentwide decisionmaking process in which public inputs must play a major role. The courts no longer deferred to the claims of the Forest Service to apply objective scientific knowledge to the management of the national forests. The power of progressive ideology to hold the fort against congressional, presidential, and other direct political interventions in the details of national forest management was being steadily eroded. By the late 1980s, groups outside professional forestry—often within the environmental movement—were setting the agenda and driving the key decisions in national forest management.

This stage had in essence been set much earlier in the world of ideas, when leading American intellectuals had concluded that progressive hopes for scientific management of American society were not going to be realized. However, unlike in Pinchot's day, the Forest Service and the forestry profession were no longer on the cutting edge of American political thought. They often would be blindsided by the developments of the 1970s and 1980s, reacting by feeling a sense of betrayal of the old progressive ideals. The legitimate place of science in guiding the management of the national forests had been usurped by private interests—the very groups that Pinchot had fought so hard to defeat. Seen from an old-fashioned progressive perspective, the forces of good government were again losing to more venal elements in American society.

ECONOMIC PROGRESSIVISM

Outside the forestry profession, leading economists who specialized in forest policy issues argued that the failure of professional forestry had been intellectual as well. If forestry was to provide a basis for scientific management of the national forests, then foresters had to do more than establish the correct rate of growth of trees under certain physical assumptions. If forestry was to be a land management science as well as a physical science, then they had to apply technical expertise to make decisions about the best use of the forests. Apparently, the "principle" of multiple use (traditionally offered by the Forest Service as the basis for allocating the resources of the forests among competing uses) could not fill this role.

At Resources for the Future (RFF), Marion Clawson, an old progressive, saw the Forest Service itself as having betrayed its early ideals. In a series of critical commentaries on agency management in the 1970s and 1980s, Clawson (1976) claimed that the Forest Service had in practice abandoned the progressive commitment to efficiency and instead was basing its decisions on crass politics. He blamed the Forest Service administrators as well as the political opportunists who sought to exploit forest resources for private gain.

Clawson's RFF colleague John Krutilla, another old progressive, offered a similar diagnosis but emphasized the steps necessary to revive scientific management (Bowes and Krutilla 1989). As Krutilla saw matters, professional foresters would have to learn economics. The only possible objective criterion for deciding among the many possible uses of the national forests was to choose the combination of uses that maximized the total value of all forest uses. Efficiency must be restored to its central role, as in early progressivism—but now, efficiency must be interpreted through the lens of economic analysis. Krutilla (1979) wrote that in this way, it would be possible to replace the "high motives and sincere exhortations" of traditional forestry with more practical "operation criteria" that could provide a true scientific basis for forest management.

The attempts of Clawson, Krutilla, and others to articulate a new intellectual foundation for scientific management of the national forests—to revive the progressive project through economics—failed as well. Economic science was not up to the scientific objectivity promised by Krutilla. Economists had no answer to the question of how the Forest Service would be able to assert managerial independence from political demands. Privatization of the forests would have been one such approach, but the progressive ethos ran toward government control.

Moreover, both Clawson and Krutilla took the progressive value system for granted. Like most forestry professionals, they automatically assumed the values of maximum use of the forests to advance human material well-being in the world, values that had in fact been at the core of Pinchot's original vision. Many members of the environmental movement, however, had a different idea. For theorists of "deep ecology," the fundamental problem was "the ultimate value judgment on which technological society rests—progress conceived as the further development and expansion of the artificial environment at the expense of the natural world" (Devall and Sessions 1985, 48). Bill Devall and George Sessions were unusually blunt and direct about their objections to the progressive value system, but similar concerns motivated many more mainstream environmentalists as well.

Renouncing scientific control over nature was an idea in basic conflict with the progressive value system. For American progressives, for example, the dam had been a great symbol of progress. It represented the wonderful application of engineering expertise to conquering a raging river to provide food and electricity for the world. The view of environmentalist David Brower, however, was, "I hate all dams, large and small," whatever the benefit–cost ratio might be (quoted in McPhee 1971, 159). Economics had nothing to do with the undesirability of building a dam because a dam was, simply put, a desecration of nature. To argue for building a dam would be like arguing for the institution of slavery because it was economically efficient. Instead of the progressive symbol of a dam, the leading symbol of modern environmentalism has been a wilderness area, defined by the very absence of human presence and impact.

The Forest Service thus found itself torn between budget officers and economists who argued for a revival of the progressive project of Pinchot, and leading environmentalists who renounced the very value system of progressivism. In the 1990s, the agency sought to reconcile these tensions by a virtual sleight of hand. A new principle of "ecological management" would put scientific management at the service of the new environmental values. But what if a basic hostility to scientific management lay at the core of much environmental thinking? What if many environmental thinkers found objectionable the very idea of manipulating nature through scientific knowledge for human benefit?

In seeking to reconcile the irreconcilable, many observers saw ecological management as empty of content—more of a public relations gesture than actual scientific management. As the U.S. General Accounting Office (GAO) put it, "ecosystem management has come to represent different things to different people." The GAO also found one person putting it cynically, "there is not enough agreement on the concept to hinder its popularity"—a comment often made earlier with respect to a previous Forest Service management "principle": the principle of multiple use (U.S. GAO 1994, 38).

Despite such problems, the Forest Service does not feel able to abandon explicitly the ideas of scientific management. It would be almost like a Christian abandoning the authority of the Holy Bible. Pinchot is the George Washington of American forestry. Moreover, scientific management resolves a very practical question: What does the Forest Service say when a forest user asks, "Why have you favored another use over my desired use?" The answer, as the Forest Service now responds, is the "scientific" dictates of ecological management. It may be mostly a bluff, but the language of ecological management at least sounds good enough to the public to buy the Forest Service some breathing space.

If there were no answer, if the Forest Service had to confess today that it has no principled basis for decisionmaking, then the social legitimacy of the agency might be in doubt. The very institutional survival of the Forest Service might be in danger. As prominent a commentator as Randal O'Toole has stated that the "Forest Service will be 100 years old in 2005—if it survives that long. There is a good chance that it won't" (O'Toole 1997).

The fundamental problem is the need to find a new guiding vision to replace the Forest Service's founding vision—its religion—of scientific management. The Forest Service is still wedded to a set of progressive ideas that have not been widely accepted among leading American political and social thinkers for a quarter of a century or more. The difficulty for the Forest Service, to be sure, is that so much of its current institutional forms and public posture has been so closely tied to scientific management themes for so long. If these arguments were abandoned, the Forest Service might have no satisfactory intellectual defense against demands for radical changes in its organization and the traditional ways of doing things. As O'Toole speculated, even the future of the agency could be in doubt. Above all, the claim to scientific objectivity has justified the centralization of authority to manage the national forests at the federal level.

It is difficult for the Forest Service to exercise a leadership role in the search for a successor vision to the ideas of scientific management, because Pinchot was a leading figure in formulating the broader tenets of

American progressivism. However, the same need not be true of the forestry profession. If scientific management were abandoned, significant changes probably would take place in the practice of professional forestry. However, as long as we have forests, we will need systematically organized inquiry into the management needs and policies for those forests.

LOOKING TO THE FUTURE

Making predictions about the management of U.S. forests in the twenty-first century is a matter of projecting both scientific and political trends. Scientific management was in essence a theory—simplistic, as it now seems—of the use of scientific knowledge in the political process. Forest management was to be divided into two tasks, supposedly to be undertaken in separate domains. Science should operate on its own terms in the distinct domain of professional expertise, and politics should operate in a separate and democratic sphere.

The failure of this political/scientific vision of scientific management clearly does not mean that science should not be part of forest decisionmaking. That could hardly be the case. Scientific knowledge is still essential for providing a base of information for forest management. However, it can no longer be assumed that the application of the scientific knowledge will be separated from the practice of politics. Instead, in the future, the use of science will be intermingled with the political process. In a commentary on the policymaking process with respect to world climate change, Brown (1996, 18) noted, "science and politics had begun to mix and would not be separated again"—an outcome that is simply inevitable.

The matter is not the difficulty in practice of separating value decisions from technical decisions. It is now apparent that science itself reflects a powerful value lens; to make decisions according to some scientific criterion will be to apply a specific set of values to the decisions. Hence, in the future, science and politics will have to be part of one seamless web of forest decisionmaking that involves technical and value elements.

In a democracy, politics must be the ultimate source of social legitimacy. Hence, what will be required is an effective way to communicate scientific knowledge to democratic political actors. These participants in the political process will then somehow have to blend scientific knowledge with the values of the communities they represent. Forest decisions will be a mixture of values and science, as brought together by the political representatives of the people.

If forest decisionmaking is conceived in this manner, it offers a strong case for the decentralization of forest decisionmaking responsibilities. The process of mixing science and values, for example, is likely to vary significantly from one local community to the next. Under the old tenets of scientific management, it could reasonably be assumed that there was one scientific answer for the whole nation. Indeed, this answer could best be determined at the federal level, through the efforts of the most skilled forestry professionals in the nation. National solutions would then be applied locally by forest administrators subject to national direction.

If values and science have to be blended in ways that are unique to each forest community, however, the justification for a national solution is much weakened. The centralization of management authority at the federal level is likely to preclude close sensitivity to local political values. Even if national administrators accurately understood these local values, the blending of values and technical considerations would be a time-consuming process. National authorities would be physically incapable of undertaking such a task for the whole country. The debate over the local forest plan of the Quincy Library Group in northern California illustrates the problem. If Congress had to undertake a similar review for every local area of the national forest system, then the entire Congress would soon be spending much of its time on forest management matters. Obviously, that is not going to happen.

Thus, if national administrators retain decisionmaking responsibility, the most likely outcome is that they will end up imposing a set of national values on local communities. However, if these communities resist, as will often be the case, then the national decisions may not hold. The likely result is what is commonly seen today on the national forests: institutional paralysis and decisionmaking gridlock.

Partly out of desperation, westerners are beginning to look more closely at institutional alternatives to the current federal land dominance. Daniel Kemmis (1990, 127–8), former speaker of the Montana House of Representatives and former mayor of Missoula, writes that the west

> cannot transcend its colonial heritage until it gains a much more substantial measure of indigenous control over its own land and resources. But it can neither gain nor exercise that control until the left and the right gain enough trust in each other, and establish a productive enough working relationship, to enable them to agree, at least roughly, on what they would seek to accomplish if they had such control.... Until that happens, ... the end of those federally controlled debates is always a less satisfying way of inhabiting the place than any of the participants would have chosen. As more and more people become dissatisfied with this less-

than-zero-sum solution of the procedural republic, it is time to look the alternative in the face.

Decentralization of national forest decisionmaking could take many forms (see Nelson 1987; 1994; 1996b). The individual units of the National Forest System could be kept within the federal system but given much greater decisionmaking autonomy. They could be given the authority to raise revenues—through user fees and other devices—to cover their costs. Thus, each individual national forest might function much like a public corporation. It would respond to local demands as expressed through local willingness to pay for various forest services and thus willingness to cover the costs of providing these services.

Another possibility would be to delegate forest decisionmaking responsibility to existing state and local political jurisdictions. National forest uses might be decided according to values expressed by representatives of the surrounding community. However, if the federal government continues to pay the way, local groups will have no incentive to economize. From a local perspective, federal money is "free" money. Thus, any balancing of money benefits and costs must require the local community to cover the costs. In short, if local values are to determine forest uses, then it may be appropriate to transfer the ownership of national forests to the state and local levels.

Yet another decentralization possibility would be to privatize parts of the national forests. An entrepreneur motivated by profit would assess the forest demands and the willingness to pay of various forest users, then allocate the uses of the forest to the parties that could offer the highest profit, that is, those that would pay the most relative to the costs of serving their needs. This is the market system by which most goods and services in American life are provided. Privatization of the national forests would work best where few external impacts of forest use are not reflected in prices in the marketplace. Such circumstances will likely arise where commercial timber harvesting or another commercial activity is the principal use of a forest, for example.

What may be needed is a comprehensive reclassification of the national forests (see Nelson 1995a; 1996a). Some lands involve matters of national significance; the uses of these lands are important in terms of not only local but also national values. Such areas would probably include much of the existing national wilderness system within national forests. These areas could remain in federal ownership.

However, many other areas of the national forests would not possess nationally significant environmental or other assets. If these areas served a variety of nonmarket purposes, the management of the areas should respond to local community values—a goal best accomplished by transferring the lands to states and to local governments. Finally, commercial lands best suited to private management also could be so categorized.

A national federal land classification commission might be formed for this purpose. It could be given a charge to systematically review all federal lands and determine which lands fall into the defined categories. It would be helpful to regularly identify specific lands that meet the criteria identified earlier. State, local, or private management of these lands might then be tried on an experimental basis. Indeed, several recent proposals by groups studying the problems of national forest management have been made along these lines (Federal Lands Task Force 1998; Forest Options Group 1998). The idea of managing lands under a trust status, similar to the current state trust lands, seems to be receiving particular attention (Souder and Fairfax 1996).

SUMMARY

The management of the national forest system derives its social legitimacy from the ideas of scientific management. Yet, the failure of scientific management theories as a political concept was apparent long ago. The basic precepts of scientific management are rejected today by leading players in national forest decisionmaking. Traditional organizational forms in the national forests and ways of operating of the Forest Service are surviving largely on institutional momentum.

We must think more boldly in the new century. New ideas to replace scientific management will be required. And these ideas will inevitably have major institutional consequences, involving brand-new alternatives to the familiar arrangements of the existing National Forest System. The Forest Service as we know it today will likely be radically altered by these developments.

REFERENCES

Biskupic, Joan. 1997. Trial Judges Have Wide Discretion On Scientific Testimony, Court Says. *Washington Post.* December 16, A2.

Bowes, Michael D., and John V. Krutilla. 1989. *Multiple-Use Management: The Economics of Public Forestlands.* Washington, DC: Resources for the Future.

Brown, Paul. 1996. *Global Warming.* London: Blanford.

Clawson, Marion. 1976. *The Economics of National Forest Management.* Washington, DC: Resources for the Future.

Devall, Bill, and George Sessions. 1985. *Deep Ecology: Living as if Nature Mattered.* Salt Lake City, UT: Peregrine Books.

Ely, Richard T. 1889. *Social Aspects of Christianity and Other Essays.* New York: Thomas Y. Crowell.

Federal Lands Task Force. 1998. *New Approaches for Managing Federal Administered Lands, A Report to the Idaho State Board of Land Commissioners.* July.

Forest Options Group. 1998. *Options for the Forest Service 2nd Century.* Oak Grove, OR: Thoreau Institute. http://www.ti.org/tc.html.

Galbraith, John Kenneth. 1956. *American Capitalism: The Concept of Countervailing Power.* Boston: Houghton Mifflin.

Hays, Samuel P. 1959. *Conservation and the Gospel of Efficiency: The Progressive Conservation Movement, 1890–1920.* Cambridge, MA: Harvard University Press.

Hopkins, Charles Howard. 1940. *The Rise of the Social Gospel in American Protestantism, 1865–1915.* New Haven, CT: Yale University Press, 320–1.

Kemmis, Daniel. 1990. *Community and the Politics of Place.* Norman: University of Oklahoma Press.

Krutilla, John. 1979. *Adaptive Responses to Forces for Change.* Presented at the Annual Meeting of the Society of American Foresters, Boston, MA, October 16.

Langston, Nancy. 1995. *Forest Dreams, Forest Nightmares.* Seattle: University of Washington Press.

Lee, Eliza Wing-yee. 1995. Political Science, Public Administration, and the Rise of the American Administrative State. *Public Administration Review* 55(6): 538.

Lindblom, Charles E. 1959. The Science of "Muddling Through." *Public Administration Review*, Spring.

Lindblom, Charles E., and David K. Cohen. 1979. *Usable Knowledge: Social Science and Social Problem Solving.* New Haven, CT: Yale University Press.

Lowi, Theodore J. 1969. *The End of Liberalism: Ideology, Policy and the Crisis of Public Authority.* New York: Norton.

McPhee, John. 1971. *Encounters with the Archdruid.* New York: Farrar, Straus, and Giroux.

Nelson, Robert H. 1983. *The Making of Federal Coal Policy.* Durham, NC: Duke University Press.

———. 1987. The Future of Federal Forest Management: Options for Use of Market Methods. In *Federal Lands Policy,* edited by Phillip O. Foss. New York: Greenwood Press.

———. 1991. *Reaching for Heaven on Earth: The Theological Meaning of Economics.* Lanham, MD: Rowman and Littlefield.

———. 1994. Government as Theatre: Towards a New Paradigm for the Public Lands. *University of Colorado Law Review* 65(2).

———. 1995a. *How to Dismantle the Interior Department.* Washington, DC: Competitive Enterprise Institute, June.

———. 1995b. *Public Lands and Private Rights: The Failure of Scientific Management.* Lanham, MD: Rowman and Littlefield.

———. 1996a. *How and Why to Transfer BLM Lands to the States.* Washington, DC: Competitive Enterprise Institute, January.

———. 1996b. End of the Progressive Era: Toward Decentralization of the Federal Lands. In *A Wolf in the Garden: The Land Rights Movement and the New Environmental Debate,* edited by Philip D. Brick and R. McGreggor Cawley. Lanham, MD: Rowman and Littlefield.

O'Toole, Randal. 1997. Expect the Forest Service To Be Slowly Emasculated. *The Seattle Times,* May 7.

Pinchot, Gifford. 1967. *The Fight for Conservation.* Seattle: University of Washington Press, 95.

Pyne, Stephen J. 1997. *World Fire: The Culture of Fire on Earth.* Seattle, WA: University of Washington Press, Henry Holt & Co.

Souder, Jon A., and Sally K. Fairfax. 1996. *State Trust Lands: History, Management and Use.* Lawrence: University Press of Kansas.

Tarlock, A. Dan. 1985. The Making of Federal Coal Policy: Lessons for Public Land Management from a Failed Program, an Essay and Review. *Natural Resources Journal* 25(2): 349–74.

Truman, David B. 1951. *The Governmental Process: Political Interests and Public Opinion.* New York: Knopf.

U.S. GAO (General Accounting Office). 1994. *Ecosystem Management: Additional Actions Needed To Adequately Test a Promising Approach.* Washington, DC: U.S. General Accounting Office, August.

Discussion

Rethinking Scientific Management

Daniel B. Botkin

THE PROBLEM

In Chapter 4, Robert Nelson eloquently and clearly defines the problem facing the Forest Service: the perceived failure of scientific management. He asserts that a belief in the power of scientific management was one of the tenets of twentieth-century progressivism in American (and Western) society. He sets forth the problem so that we can address the subsequent issue: What can we do about it?

Science has not so much failed as it has never been correctly applied. Therefore, part of the answer is to apply science correctly. Only the perception of it has been put forward. (Note that the failure to apply science correctly extends to all kinds of management of natural resources.)

RESTATING THE PROBLEM

Science has never been properly applied and therefore never has been given a chance. The failure to apply science is partly the result of the belief in steady-state, balance-of-nature ecological systems.

Scientific management of U.S. forests was not attempted for many reasons. As I discussed in *Discordant Harmonies: A New Ecology for the 21st Century* (1990), the management of forests and other biological resources was not scientific in the sense of a direct application of validated scientific concepts or as part of management that involved experimentation

---

DANIEL B. BOTKIN is a professor of ecology at George Mason University in Virginia and president of the Center for the Study of the Environment, which is based in Santa Barbara, California.

that could be used to demonstrate or test the scientific basis. Instead, a set of beliefs was accepted as true and *scientifically based* when those beliefs were based on ancient, in fact, "prescientific" concepts derived from the great myth of the balance of nature. These beliefs are contradicted by almost all available scientific information when that information is carefully examined.

According to the myth of the balance of nature, a forest, like any other ecological system, achieves a permanency of form and structure that persists indefinitely if left undisturbed by human influence. When a forest is subjected to disturbing agents but then freed from them, it returns to its original fixed (steady-state) condition. In the twentieth century, the balance-of-nature myth was recast as a belief in a natural steady state—in standard statements about the resistance and resilience of ecological systems, in the idea of ecological succession to a single climatic climax for a forest type, and in mathematical equations used to calculate forest yield. For example, the site index used by foresters to calculate growth and yield from a stand assumes that a particular location is capable of producing a tree of a certain size, independent of (that is, making no allowances for) variations in the environment or the biota. These tenets of twentieth-century forest management demonstrate that the balance of nature was so strongly believed that it was assumed to be a scientific truth without being tested.

Lack of Scientific Data

Scientific management requires scientific data, but adequate data about basic and fundamental properties of a forest, monitored over time, never have been obtained. To determine whether forest practices lead to sustainable production, it is necessary to measure the diameter and height of individual trees on sample plots, record the species of each tree, and measure certain soil properties—preferably, all at regular intervals and just before harvest—and then record the amount of timber removed from these plots during harvest. Some plots must be maintained as controls (that is, not harvested), for which the same data are measured.

Although the Forest Service has a system of permanent plots, I have searched for thirty years to find a single control plot that was measured twice by using the same methods. I failed to find any; however, every time I bring up the issue in a speech, someone from industry or a government agency comes up to tell me that the records exist. They tell me whom to call. Invariably, that call leads to another, which leads to yet another, in a seemingly endless chain of calls—until finally, I contact someone who suggests I call the person I called first. This circle leads nowhere.

It is likely that somewhere in the Forest Service's files, data exist for a few plots that were so measured. But these records were the result of one person's commitment to the importance of monitoring, rather than the institution's commitment. Therefore, the results reside with that individual, rather than the organization. I was told that one such set of measures was completed by a Forest Service scientist who is now retired. Unless a current employee becomes interested in that set of data and works with the retired scientist, the records eventually will be lost. The process of obtaining the correct kind of data, maintaining it, analyzing it, and applying it to management is not institutionalized; it is localized.

An Institutional Failure

The failure to monitor necessary variables over time is partly a result of the widespread belief in the balance of nature. If a system must return to a fixed steady state and that steady state is known from theory, then the perception of the managers and scientists is that no measurements need to be made. The desired amount of harvest and harvest rotation can be calculated from simple formulas. However, if the environment varies and biota change over time (for example, species migrate in or out), then the system is not steady state, and monitoring is necessary. *Without such monitoring data, scientific management is not possible.*

Prejudice against Measurements

Monitoring the same variable at periodic intervals for forests has been considered "nonsexy" research or not research at all by most forest scientists in the twentieth century. Therefore, most scientists have avoided the task. Academic scientists have tended to view long-term monitoring as applied research, and applied research has a lower status than so-called fundamental research. Scientists also have looked down on monitoring that was not explicitly connected to a single hypothesis, even though it might have wide-ranging importance. The long time required for forest monitoring exceeds the professional lifetime of an individual scientist and therefore has been seen as unproductive for professional advancement.

The measurement of species, height, and diameter of trees at ten-year intervals suffers from all these prejudices. Although attitudes have begun to change during the past two decades, the effort is still small and not sufficiently institutionalized across the Forest Service. This failure extends worldwide and is a symptom of our times; the problem is not unique to the Forest Service.

Integrating Management and Science

Although among federal agencies the Forest Service has had some of the best scientists trained in the study of a biological resource, the relationship between management and scientific research has been generally disconnected. The fault has been on both sides, the scientists and the managers. In an ideal situation, scientific management involves what is referred to today as "adaptive management." Goals are clearly defined, and scientific research is applied to determine how to best achieve those goals. Initial policies are set from existing available information but are flexible and adjusted as new scientific information arises. This idea has not been prevalent in the United States.

Forest managers have not always respected or paid attention to the work of forest scientists, nor have they worked closely with those scientists. Instead, managers tended to operate under the belief that they understood the forest systems from their personal field experience and knew how to manage them without much additional input from the scientists.

Creative scientists, however, tend to want to follow their own paths and interests. Therefore, much of the research done by Forest Service scientists has departed from the overarching goals of the organization and focused instead on specific research problems of interest to the researchers. Once again, among scientists, those who progressed on their own, on problems of their own choosing, have been recognized as being of higher status than those who work in multidisciplinary teams on a societal goal. This problem is not limited to the study of forests; colleagues in cancer research tell me their field suffers similarly. Part of the difficulty is that career advancement tends to be based more on individual scientific accomplishment than on group accomplishment.

Also, Forest Service scientists have been pulled away from field work because of other demands on their time, such as frequent calls to testify as scientific experts in disputes over logging in specific areas. These activities have detracted from opportunities to focus research on larger issues and to communicate and work with managers, if they had wanted to.

Perhaps more fundamentally, the world view or mind-set of a direct connection between management policy and scientific research has not been strong in the United States with any of its biological resources. Typically, managers have not received active day-to-day interaction with their scientists, nor have they received timely direct feedback of the results of scientific field studies to help them make informed decisions.

This disconnect between management and science is perhaps best illustrated by a comparison of the approaches to the management of biological resources in the United States and management of wildlife in national parks of eastern and southern Africa. In many African nations, scientists and managers have worked closely together to try to sustain wildlife for public viewing. For example, in Zimbabwe, Sengwa is a national park devoted to scientific research concerning wildlife management, and at Krueger National Park in South Africa, the managers and the scientists have been closely linked. Perhaps the close connection between scientists and managers in these African parks has developed partly because mistakes in wildlife management become quickly evident, whereas mistakes in managing forests usually are expressed over longer periods.

Forest Dynamics versus Science

Compared with the lifetime of a tree and the time involved in forest regeneration, the Forest Service is young. Its total existence is shorter than the longevity of many major tree species, less than two or three harvest rotations of many commercially valuable species. The same can be said of ecological and forest sciences in an academic setting; they also are young in terms of data collection. In general, the sciences related to the management of natural resources are young in a conceptual sense as well as a temporal sense, and the topic is complex. Only within the past several decades have some key aspects of these sciences developed.

As a result, the scientific methodologies that help bring scientists closer to management and applied issues—for example, geographic information systems—are only now becoming popular among ecological and forest scientists. Under the rubric of adaptive and ecosystem management, the use of these new methods emphasizes the kind of communication that has been lacking between management and scientists and the kind of connection that has been lacking between management goals and scientific methodology. It is ironic that just as methodologies are becoming available and attitudes are changing among scientists and managers, scientific management might be eliminated from the Forest Service on the grounds that it has failed.

Scientific management has been used in name only. It is also ironic, at a time when the public arguably has more interests and concern with forests than ever before—partly because of the potential role of forests in sequestering carbon dioxide to help slow global warming, partly because of growing public concern with biological diversity attached to forests, and partly because of the growing recognition of the recreational potential of forests—that there would be a strong movement to eliminate the scientific management in the Forest Service. This possibility is especially disconcerting because the Forest Service is the only federal agency primarily concerned with biological resources that has its own internal scientific research staff. (All the researchers in the Department of Interior—which includes the National Park Service, the U.S. Fish and Wildlife Service, and the Bureau of Land Management—have been moved to the U.S. Geological Survey.) If management and applied scientific research have been disconnected in the past, then the solution would not seem to lie in separating the two camps by this additional administrative distance.

Science and Scientists in Society

Nelson suggests that the early twentieth-century faith in science was so great that science was expected to provide the end, not only the means. That science can tell society what to do is a fundamental misunderstanding of the roles of science and scientists in a democracy. Scientists in our society often are called upon to set goals. "What should we do with our forests?" is asked in terms of not only the techniques for successful management but also the end products that we should accept.

Because science deals with how and what—not why—this question should not be posed to scientists. Scientists can tell society what goals are possible and present various choices for achieving those goals. If we wanted a Socratic government of philosopher-kings, then we would ask our scientists to make policy, but such are not the makings of a democracy. And asking scientists to double as politicians likely would lead to failure.

The path to understanding the role of science in the Forest Service is to understand the societal roles of science, management, and citizens with these issues. In a democracy, the citizens choose the goals. But with a complex issue that involves ecosystem dynamics and population dynamics, the nonexpert public cannot determine possible goals. Scientific studies show which goals are possible. After the public has chosen a goal, scientists can explain how to achieve that goal technically, and what is gained and lost in obtaining that goal. This is the applied professional role of science that should be given a chance with the Forest Service.

LOCAL AND SCIENTIFIC KNOWLEDGE

Nelson points out that past practitioners of science have dismissed local knowledge grounded in practical experience. I believe that this has been true and that it has been a mistake. Local knowledge often can bring to these issues an interest free of ideologies (such as the belief in the balance of nature) that impede the development and application of science. Local knowledge is often valuable for detailed understanding. Although informal local knowledge cannot replace scientific tests, it can provide a valuable starting point, an insight to be followed by careful scientific research to demonstrate whether the assertions were correct. Thus, another part of a new paradigm for the Forest Service is a kind of balance of powers among kinds of knowledge, wherein local knowledge is a vital member of the discussion and made available as possible insights, then subjected to scientific test. Without these tests, our society has fallen victim to what I have referred to as solutions by plausibility: "If it sounds good, it must be true." Millions—indeed billions—of dollars have been wasted on these so-called solutions. One case in point is the management of salmon and their forested habitats in the Pacific Northwest, for which I have documented numerous stories (Sobel and Botkin 1995).

Environmental issues including the use of the nation's forests have been perceived as a search for the single truth about these resources. Instead, part of the new paradigm is to understand that these issues are more of design than of truth. Given a set of societal goals for the national forests, we can achieve them in many ways, just as there are many ways to build a bridge where society decides one is needed. Viewing our national forests as a question of the most appropriate design, rather than a search for the single truth, would be a major improvement. The analogy of design contents for public landscape architecture and buildings could work here. For example, in the development of a new park or a new bridge, companies often submit designs in competition. First, a professional committee selects the best of all the designs. Then, this set is made available for public comment and public choice.

One would hope for a design-competition approach to national forests, nationally or individually, by forest. Scientific review of existing data could provide knowledge of possible goals. These goals could be made available to the public, who could select (through voting) a set of goals. These goals then would be open to a design competition, which would follow the process I just outlined. In this way, scientists and citizens would function in their appropriate roles, science would be given its appropriate chance and test, and our natural resources would be much more likely to persist and become sustainable. Critics may view this discussion as merely a continuation of the belief in the gospel of efficiency. However, it does not claim efficiency as the goal; it sets the use of science within its proper context.

SEEKING A MISSION FOR THE TWENTY-FIRST CENTURY

Nelson writes that the Forest Service "lacks a clear sense of direction and mission." I believe that this statement is true at the national level and in the perception of many nongovernmental organizations, citizens, and perhaps people in industry. I have presented a case that attributes this situation to both a technical failure to achieve objectives (because of the failure to apply science) and a societal failure that resulted from society's misunderstanding of the role of science and scientists. The development of a Forest Service mission requires contact with local knowledge and concerns as well as the integration of management and science.

In the future, the success of the Forest Service—and any natural resource agency—will require the proper application of science embedded within a realistic perception of the roles of science and scientists in the management of biological resources. The capability is within our grasp. We lack only the context and perception required for the successful application of science and therefore a renewed vigor for the Forest Service.

REFERENCES

Botkin, Daniel B. 1990. *Discordant Harmonies: A New Ecology for the 21st Century.* New York: Oxford University Press.

Sobel, Matthew J., and Daniel B. Botkin. 1995. *Status and Future of Salmon of Western Oregon and Northern California: Forecasting Spring Chinook Runs.* Santa Barbara, CA: Center for the Study of the Environment.

5

Forestry in the New Millennium

Creating a Vision That Fits

Clark S. Binkley

As the 1900s drew to a close, it was fashionable to speak solemnly about the "next millennium," implying that timeless truths lurked just beyond the shadows. Well, now we are face-to-face with the reality of a new year, a new century, a new millennium—and the need to create a new vision for the Forest Service. The kinds of changes outlined here will play out over the next several decades. In particular, continued increases in the value of the environmental services that forests provide will pose great difficulty for modes of timber production that involve natural forest management and, ironically, for the preservation of environmental values on those lands dedicated primarily to that purpose.

Forestry is an ancient profession, with long traditions in Asian as well as European cultures. Before looking to the future, let us begin with the past—the first written English definition of forests. In Johnson's dictionary, published in 1701, "forest" was defined as "[a] certain territory of woody ground and fruitful pastures, privileged for wild beasts and fowl of forests, chase and warren to rest and abide in, in the safe protection of the King for his pleasure." By implication, this definition also defined forestry. The forester's sworn duty was to protect the "vert and venison" (that is, the wildlife habitat and the wildlife itself). Failure to do so was punishable by flogging or death.

CLARK S. BINKLEY is senior vice president of investment strategy and research for Hancock Timber Resource Group in Boston, Massachusetts, and former dean of the Faculty of Forestry, University of British Columbia. His e-mail address is cbinkley@htrg.com.

Two points relevant to our present circumstances stem from this definition. First, forests, in their totality, must accommodate a wide variety of users. Forests where human interference is absent (or nearly so) have important spiritual, economic, ecological, and cultural value. Forests help regulate key global biogeochemical cycles. They provide inspiration to poets and artists. For billions of people, forests form the backdrop to everyday life. Wood is one of the world's most important and most ubiquitous sources of energy and industrial raw material. In addition, millions of aboriginal people call the forests "home." Second, the word "forest" is not so much a scientific term as a social construction, dependent on contemporary norms and values. That we chuckle at the archaic language and meanings of three centuries ago simply confirms this point.

Before we think about the future of forest management, we should consider how we arrived where we are today. In the first section of this paper, I tell the story of the depletion of old-growth timber and the transition of the forest sector to reliance on forests created by human stewardship. This transition has several decades to run, so understanding it provides a helpful context for evaluating the near-term performance of alternative forest management approaches. In the second section, I discuss three alternatives (natural forest management, plantation forestry, and the management of parks and preserves) that respond more or less well to the underlying economic and social trends. Conclusions reflect on the enduring responsibilities of foresters and the special problems that the Forest Service faces.

IN THE LONG SWEEP OF HISTORY

Because of the long production cycle for forests and the large standing inventory of trees relative to annual usage, the forest sector possesses a degree of temporal momentum that is present in few other aspects of human endeavor. This momentum permits us to foresee the outline of developments over the next several decades, even if the details are unclear and uncertain. To understand the dynamics of forest sector development, let us consider forests at the global scale in the long sweep of history. I take the economist's perspective of holding everything else constant, recognizing that the real world harbors a rich fabric of complications.

The underlying story is the transition from a "hunter-gatherer" phase of forestry to a "husbandry-stewardship" phase. For agriculture, this transition began 10,000 years ago, when our ancestors discovered einkorn wheat growing wild in the Fertile Crescent (Diamond 1997; Huen and others 1997). Prior to that discovery, subsistence required humans to glean sustenance from numerous wild organisms. Now we get almost all of our food from about ten domesticated plants and five domesticated animals. In developed countries, a few of us still scavenge from the wild Earth, but only for sport, hobby, or other atavistic reasons. The wild landscape is the source of only a tiny fraction of our food, and for agriculture, the transition is now complete. Because the production cycle for trees is comparatively long and the inventory of standing timber is large relative to annual consumption, the transition in our use of forests will take far longer than it did for agriculture. This difference in transition time explains much of our circumstance today.

Timber depletion drives the dynamics of the transition. Early in the transition, timber prices are low and forest land is more valuable for other uses, especially for food production, so the trees removed are not replaced. In an old-growth forest, net accumulation of merchantable timber is low or negative (especially on the economic margin), so any harvest at all causes the standing inventory to decline. As the timber inventory declines, timber becomes increasingly scarce. Timber prices rise for two reasons. First, as logging moves into more distant and more difficult terrain (the outward shift of the extensive margin), increased logging costs push prices up. Second, because of the connection between the markets for timber and the markets for other capital assets, timber rents increase (Lyon 1981).

Three factors limit the increase in timber prices. First, as prices in one region increase, it becomes economic to operate in other regions that harbor primary forests. For example, the relatively high timber prices in the United States in the late 1960s and early 1970s opened the door for expanded production into British Columbia. Similarly, high prices in British Columbia opened the door for the Russian Far East to expand production into Asian markets.

Such interregional shifts in the extensive margin are only a finite response to timber scarcity. As prices rise for timber from natural forests, purposeful husbandry of planted forests becomes economic. Examples abound worldwide—in almost all the currently managed forests in Europe; some of the land in the U.S. South and West, Australia, and Brazil; and much of the forest in New Zealand and Chile. A recent comprehensive and definitive study found the median cost of timber production (including land rental, silvicultural costs, and interest) to be US$7/m^3 for teak, $8/m^3 for eucalyptus, and $20/m^3 for southern pines (Neilson 1997). These costs are far below current stumpage prices for natural forests in much of the world.

Finally, increased timber prices drive numerous changes in how we use wood. On the supply side, high timber costs lead to conservation in its use in manufacturing. For example, during the past decade, the lumber recovery factor (that is, the amount of lumber produced per unit of log input) has increased 1.4% per year, despite deterioration in log quality (Binkley 1993). On the demand side, engineered wood products that use wood more efficiently (oriented-strand board, medium-density fiberboard, wooden I-beams, Parallam, and so forth) have become more prominent. For example, wooden I-beams now command 25–30% of the North American floor joist market, displacing wide dimension (2 × 10 and 2 × 12) sawn lumber. More generally, each 1% increase in lumber price leads to a 0.3% increase in steel consumption, a 0.15% increase in cement consumption, and an 0.65% increase in brick consumption (Binkley 1993). These alternate materials are increasingly used *despite* their higher environmental cost.

The second key element of forest sector development relates to the increased demand for the environmental services of forests. The demand for most of these services is highly correlated with personal income. The evidence for this connection is well-documented for some aspects of the environment (for example, clean air, clean water, and outdoor recreation), and the positive relationship between income and environmental value is probably true overall.

That humans value the aesthetic aspects of the natural world more after their material needs have been satisfied is not surprising. In an excellent review of human relations with forests in the antiquity, Perlin (1991, 120) commented,

> Seneca articulated the romantic view of forests shared by many of the leisure class of his time: "If you ever have come upon a grove that is full of ancient trees which have grown to unusual height, shutting out the view of the sky by a veil of pleated and intertwining branches, then the loftiness of the forest, the seclusion of the spot and the thick, unbroken shade on the midst of open space will prove to you the presence of God."

One cannot help but note the similarity of this comment, made nearly two millennia ago, to contemporary descriptions of old-growth forests.

However, most environmental services operate outside of formal markets either because they are true public goods (that is, consumption by one person does not diminish their availability for others—aesthetically pleasing landscapes are a good example) or because society has chosen not to allocate them via markets (for example, domestic water supply and outdoor recreation). It is a simple truism that such goods are systematically overconsumed and underproduced. Market-based patterns of use do not and cannot reflect the societal values of these inputs and outputs. Forest sector development may decrease the supply of these environmental services. Increased income (ironically, partly created by the exploitation of forests) will increase the demand for these services. Prices are not available to signal relative scarcity and to induce the socially appropriate changes in production and consumption. When the mismatch between the supply and demand for the environmental services from natural forests grows large enough, governments will intervene through forest practice regulations and land set-asides. Both kinds of intervention will push the costs of timber production still higher.

These predictable trends have three important implications for forestry in the new millennium:

- Price increases for timber are limited by the availability of plantation technologies to grow industrial roundwood and by our capacity to substitute wood-saving technologies and other materials for traditional wood products.
- The implicit "price" of environmental goods and services provided by the forest is not limited. Increased income has no limits, and few, if any, technical substitutes exist for these environmental services.
- Forest land has become scarce; many demands compete for forest land. Therefore, it is logical to substitute other factors of production—especially capital and technology—for land, particularly for industrial timber production.

The challenge for the forestry profession is to craft forest management approaches that respond to these overarching economic imperatives. Three general approaches appear to be available: natural forest management, plantation-based technologies, and nature preserves. In the remainder of this chapter, I discuss each of these options.

NATURAL FOREST MANAGEMENT

The key challenge in natural forest management is to produce industrial timber at an acceptable cost while protecting critical environmental value on precisely the same areas used for timber production. This objective usually is accomplished through management by a public "integrated resource management" agency that operates under some kind of forest practice code, perhaps with forest certification added on. This forest management technology faces great difficulty, and I am skeptical about its success for five reasons.

First, mixing market and nonmarket "products" creates incompatible objectives for a public agency. The costs of each product cannot be measured unambiguously, either in theory or in practice. For example, the timber inventory comprises both the "factory" for producing timber and wildlife habitat and the vegetation needed for watershed protection.

There is no unambiguous way to ascribe the cost of creating and maintaining the timber inventory allocated to each output. Furthermore, the measurement of the ecological outcomes of alternative management regimes is difficult, especially for an increasingly skeptical public. Commenting on ecosystem management, Alverson and others (1994, 159) noted,

> These approaches imply that we know, or will know in the near future, how to actively manipulate forest stands for timber production while sustaining all valid aspects of biodiversity on the same acreage. Because [such] claims ... have been shown to be invalid or overly optimistic once they were seriously evaluated scientifically, we urge readers to carefully evaluate similar claims made in this present context.

Remember Gordon Baskerville's famous forest management dictum: "If you can't measure it, you can't manage it." In the absence of good measures of input costs and product outcomes, public agency managers probably cannot appropriately balance—at the same time and place—all possible outputs of a forest.

Second, forest practice codes are costly, with no guarantee of positive ecological outcomes. For example, the recently enacted British Columbia (B.C.) Forest Practice Code reduces the size of cutblocks and imposes minimum green-up ages before adjacent cut blocks may be harvested. These prescriptions sound positive, or at least harmless—or so it would seem. But detailed simulations of the long-term consequences of these simple rules depict high degrees of forest fragmentation, reduced forest-interior habitat and increased forest-edge habitat, and a great increase in the length of the active road system. All of these effects produce negative environmental consequences.

Whatever the ecological effects, these changes in forest management greatly increase the cost of industrial timber extraction. In 1992, coast-wide logging costs totaled \$57/m^3. By 1996 (after the forest practice code was fully implemented), costs had increased more than 50% to \$88/m^3. Largely as a consequence of these increased costs, the return on capital used in the B.C. forest industry is now \$700 million/year less than its cost of capital. The capital stock is not being renewed, and the industry will become increasingly noncompetitive and unsustainable.

Third, forest certification—perhaps the endpoint of forest practice regulation—does not appear to be the answer. Aside from process issues related to the Forest Stewardship Council criteria (which are largely arbitrary, and the development process excludes key stakeholders), widespread adoption would appear to severely curtail the use of wood, one of the world's only renewable sources of industrial raw materials and energy. In a recent definitive census of all forest certified at the time (about 0.1% of the world's forests), Ghazali and Simula (1996) found that the average output of industrial roundwood on certified forests was 0.7 m^3/hectare/year. These data imply that the world would require about 4.7 billion hectares of forest to meet annual demand for wood and an increasingly larger area each year as demand for wood increases proportionally with population and wealth. Unfortunately, the world's forests now amount to only 3.6 billion hectares! Hence, managing all forests to the low level implied by the current Forest Stewardship Council standards would condemn all wild forests to industrial exploitation by humans.

Finally, while forest management costs increase, public revenues will fall. Because the value of nonmarket environmental goods and services is increasing relative to timber values, it is logical to shift the production mix away from timber toward these nonmarketed goods and services. However, the budget for public forest management agencies usually is proportional to cash income, either implicitly (through the political budget process) or explicitly (if some agency revenues are held internally). So, we can infer that the resources available for forest stewardship will decline just at the time that the forest management problems become more complex. The Forest Service has eliminated thousands of jobs as the timber program has declined. People interested in the nontimber aspects of forests typically do not lobby for increased budgets for the integrated resource management agencies. Consequently, we can justifiably conclude that the natural forest management paradigm is not sustainable because it does not (and probably cannot) provide the resources necessary for responsible forest stewardship.

In short, the natural forest management paradigm that has dominated the practice of forestry since its inception faces daunting challenges. The information base required to retain public confidence has exploded just as forestry management and research budgets are declining. The regulatory structures that are imposed to deal with the increasingly valuable environmental outputs of our forests have unknown, and possibly negative, environmental consequences—but they do have high costs. These costs are registered in a declining forest products industry and a decline in the use of one of the world's most environmentally friendly materials and energy sources. The history of public forest management agencies shows us that budgets will decline as the politicians responsibly and appropriately shift the national forest output mix away from the traditional emphasis on timber production. Because of these difficulties, I predict a continued struggle and the early demise of natural forest management.

If natural forest management is not the answer, then where will we get our wood?

PLANTATION MANAGEMENT

Plantations permit foresters to substitute capital and technology for land. This method is an extraordinarily powerful tool.

Studies initiated by John Gordon in the early 1970s examined the maximum theoretical timber yields based on the biochemical efficiency of trees in turning sunlight, water, and carbon dioxide into economically usable plant parts. The Weyerhaeuser Company applied these models to study sites in the U.S. South and Pacific Northwest, where timber culture is most intensive. Yet on these sites, the best management practices achieved only 40–50% of the modeled maximum yields and natural stands on the sites produced 10–25% (Farnum and others 1983). Across the globe, specific timber yield increases of two- to fivefold appear to be generally technically feasible and economically attractive.

Plantations are appealing because of this large potential to free natural forests from intensive exploitation for industrial purposes (see Sedjo and Botkin 1997). Within a region, every one hectare of plantation forest can free up to five hectares of natural forest from industrial timber production. The substitution is far greater between regions; for example, one hectare in Brazil can replace perhaps twenty hectares of land in Siberia.

Combined with sophisticated wood products technology, plantation-grown wood can substitute for most—if not all—of the products obtained from natural forests. Indeed, the uniform (and possibly designed) fiber characteristics of plantation-grown wood make it more desirable for many products. The use of sophisticated engineering concepts and small amounts of nonwood materials in such products as laminated veneer lumber or oriented strand lumber will obviate the need for traditional sawlogs.

If future timber production is focused on plantations, then how and where will we provide for the key nontimber values of forests?

PARKS AND PROTECTED AREAS

Parks and protected areas will become increasingly important elements in the forestry equation. But however challenging the situation is for natural forest management, the circumstances are far more difficult for protected areas. Establishing parks and protected areas creates at least three kinds of problems: philosophic, managerial, and technologic.

The philosophic problems arise directly from our name for these lands: "protected areas." Protected from what? Presumably, the answer is, "protected from human interference." Hence, we define "nature" tautologically as "forests in the absence of humans." This tautology, as

Cronon (1996, 81) noted, is extremely problematic: "[I]f nature dies because we enter it, then the only way to save nature is to kill ourselves. The absurdity of this proposition flows from the underlying dualism it expresses.... It is not a proposition that seems likely to produce very positive or practical results."

One popular standard of nature, so defined, is those conditions existing prior to European contact. Such a definition is deeply troubling. In the first place, this definition is a profoundly racist form of neocolonialism. If nature is defined as the forest without humans, then one implication is, of course, that indigenous peoples are not human. This thinking is nothing less than another form of *terra nullius* that has been the basis for European subjugation of native people throughout North America.

Second, this definition ignores the profound impact of indigenous peoples on forests. Although the literature is now growing on this point—with excellent examples from Europe, Asia, South America, and Oceania—let me quote just one example from North America. Pyne (1997) noted,

> In 1878 John Wesley Powell published a map of Utah for his *Report on the Lands of the Arid Region of the United States*, long celebrated as a seminal document in the history of conservation. His crews had classified land by four categories—desert, irrigable, forest and burned. The burned lands proved the largest in area, rising from the grasses that fringed the desert to the rims of the high plateaus. The primary source of fire was the local Indian tribes, burning for their traditional reasons. Given time (and not too much of it) the burned area would overwhelm the forest, and with its watershed ruined, the irrigable would regress into desert. The future of the Rocky Mountains, Powell lectured, depended on the stability of their forested watersheds, and the future of those forests depended on fire protection.

Instructively, the title of Pyne's chapter is "The Great Barbecue."

The Wisconsin glaciation apparently forced Native Americans southward. As the ice sheet retreated, people migrated north, living in the productive successional forests at the edges of the glaciers. Consequently, people occupied parts of North America before trees did. In such areas, we can say precisely that we have no knowledge of ecosystems in the absence of humans.

This philosophic problem implies the management problem. If we (justifiably) reject pre-European settlement conditions as the ruling definition of nature, then we must purposefully choose the specific ecosystem conditions we want to sustain. Proctor (1996, 295), reflecting on what he calls the "contested moral terrain of ancient forests" in Oregon and Washington, puts the question in its postmodernist form: "Whose nature should we save?" He answers, "There is no one nature to save in the Pacific Northwest since nature is always in part a social construction." Management of parks and preserves requires thoughtfully disentangling our social constructions of nature to separate the elements that arise from values from those that arise from objective features of the landscape (Binkley 1998).

An example from British Columbia will illustrate the point. The 45,000-hectare Khutzymateen, a spectacular drainage on the north coast adjacent to southeast Alaska, was declared as a grizzly bear reserve in 1994. Grizzlies are abundant in the valley, largely, I am told, because of the beneficial mix of early and late successional habitat in close proximity (also because the valley is inaccessible to public roads, so hunting pressure—legal or illegal—is nil). The early successional habitat is the result of logging and, in the absence of further disturbance, will disappear over time. Our current construction of the term "park" precludes precisely the activities needed to create this kind of habitat. The population of bears will likely decline as a result, and we anticipate the tragic irony of a grizzly bear reserve that contains few of its namesake species.

The effects of technology on conservation are poorly understood but follow quite logically from the management problem. In the Khutzymateen, for example, how can we maintain a high-quality bear habitat in the absence of logging? And how can we do so at an acceptable cost? Such questions demand a substantial, focused research program, and foresters are well qualified to lead it.

CRAFTING A SUSTAINABLE VISION

In the new millennium, the role of foresters will remain much the same as when Johnson penned his definition in 1701: stewardship and husbandry of forest lands for the wide range of values they provide. Only the economic context, our knowledge of the underlying production technologies, and our social constructions of "What is a forest?" have changed.

It is logical and appropriate for the forestry profession to respond and adapt to these new circumstances. Natural forest management is contentious and will decline in importance as a source of industrial timber supply. Because of the large standing inventory of timber and the significant amount of installed industrial capacity that relies on this management approach, the end game will take quite a long time. As timber production declines in importance as part of this paradigm, the resources available for forest stewardship will decline, and management standards will deteriorate. Might it be wise public policy to accelerate the demise of this outdated, unsustainable management approach?

For industrial timber production, intensively managed plantations are becoming increasingly important. This trend is especially true on lands currently classified as nonforest—marginal agricultural lands, pasture lands, and grasslands. With all the political attention directed to the declining years of natural forest management, are we adequately investing in education, research, and technology related to plantation management? From my recent experience as dean of the Faculty of Forestry at the University of British Columbia, I know that in Canada, we are not.

Parks and preserves represent a major opportunity for contemporary forestry and to re-embrace the "vert and venison," the concern for which the profession was initially founded.

One major impediment to effective public forest management is understanding public desires for alternative approaches. After conducting thousands of interviews, the social psychologist Stephen Kellert (1993) identified a typology of nine different attitudes, or "constructions," of nature. Whereas the structure of the typology appears to be robust to cultural differences, the frequency distribution of individuals among the various categories differs dramatically among countries. For example, Kellert (1995, 110) found "a Japanese public far more inclined than the American to emphasize control over nature." The North American concept of wilderness is wholly absent in Germany, where humans play a leading role in the landscape. Some advocates would have us choose one construction of nature over all others, but enforcing such a choice is inconsistent with the liberal notions that underlie Western democracies.

The key to successful pluralistic forest management, then, is to develop management technologies that correspond to the various social constructions. Although they are well-developed for utilitarian and dominionistic constructions of nature (to use Kellert's typology), they are comparatively poorly understood for naturalistic or ecologistic constructions.

The Forest Service would logically respond to this diversity of constructions of nature by not choosing one approach but rather crafting a mix of these approaches to apply to different lands under its management. Each approach has its own relevance to the peculiar institutional and ecological characteristics of the national forests.

Some national forest lands are suitable for intensive timber management. The fraction is probably small and is located primarily in the U.S. South and Pacific Northwest. Despite the innate suitability of these lands for plantation management, the Forest Service has no particular comparative advantage in applying this management regime, for two reasons. First, plantation forestry is among the most capital intensive activities known. Public agencies do not reward profitable deployment of capital, so it is unlikely that the public sector can be as effective as the private sector in plantation management. Second, almost all of the benefits of plantation forests flow through the market, so the public sector has no particular reason to engage in this kind of forest management. Therefore, it would be logical for the Forest Service to spin off to the private sector, through trades or outright sales, its lands that are particularly suitable for intensive management. Any funds raised in this manner could be redeployed either in a trust fund to support nontimber aspects of existing forests or, alternatively, to purchase private lands (or land use rights) that provide substantial public values.

Because the Forest Service already looks after extensive areas of land—the national wilderness system—for nonindustrial purposes, managing forest lands as nature preserves is clearly consistent with its mission. Surely, the preserved areas within the national forests will increase over time. However, as I argued earlier, this form of management still requires the articulation of the ecosystem conditions to be sustained. As Kimmins (1994) pointed out, the currently popular notion of "ecologically sound" management alone excludes very little until a particular ecosystem state is declared preferable to other possible ones. To develop and sustain a leadership position in preserve management, the Forest Service will have to (1) develop effective means of discerning socially preferable ecosystem conditions in the face of chaotic, pluralistic interactions among stakeholders, each of whom holds particular—and possibly mutually exclusive—social constructions of nature, and (2) create and apply new technologies for sustaining the chosen conditions. As the total national forest land in preserve management increases, so the need for public subsidy of the Forest Service also will increase. If the needed resources are not forthcoming, stewardship standards inevitably will fall.

Finally, the national forests may have a unique opportunity to craft a sustainable version of the natural forest management paradigm, but major organizational and institutional changes would be required. Of specific concern is the need to maintain adequate resources for excellent stewardship in the absence of a large timber program. One approach would be to make the national forests a corporation instead of a government organization, along the lines of the New Zealand Ministry of Forests (prior to privatization) or the U.S. Postal Service. Governed by an elected board of directors, the "U.S. National Forests Corporation" would be empowered to sell *all* the products of the forest, from water to wood, with a mandate to maximize the asset value of the land base under the constraint of annual audits of public stewardship. Such an organization would pose no great difficulty for products currently traded in markets (for example, timber) or for products that could readily be marketed (for example, recreation and water). For true public goods such as aesthetically pleasing landscapes and carbon fixation, the corporation would contract with the U.S. Congress or others to provide the needed services. The public interested in the increased provision of such goods would lobby the U.S. Congress to increase the contract rates. The corporation might securitize certain income streams to raise needed capital. Only through such an organizational mechanism might adequate funds be available for forest stewardship.

Such a change also would imply changes in Forest Service research activities. The research program logically falls into two categories. One responds to the public good aspects of forest-sector research and development (R&D) related to the forest industry. Because it is difficult for any one firm to fully appropriate all the benefits of new innovations, individual firms will systematically underinvest in R&D. Along with the nation's universities, the Forest Service's research program is one means to respond to this situation. The returns to forest sector R&D have been high (Hyde and others 1992) and are apt to remain so as long as the United States is a net importer of wood (Binkley 1997). The public interest in forest-sector R&D could be met in several alternative ways, including direct public subsidy (as is now the case) or via an R&D tax levied on the consumption of forest products.

The second part of the research activity responds directly to the R&D needed to support national forest management. There is no reason to suspect that the optimal research program for this purpose would bear any relationship whatsoever to the one needed to respond to the public good aspects of forest-sector R&D mentioned earlier. The corporation might contract with a national forestry research body or with universities, or it might have its own in-house research facility. Confusing these two general purposes of national forest research serves neither well.

Foresters often lament the long periods involved in forest management. Ironically, precisely this characteristic of the sector permits us to predict with some certainty the major underlying trends that characterize forestry in our era—the transition, for industrial purposes, from reliance on old-growth forests provided wholly by nature to second-growth forests created through human stewardship. Organizations well-adapted for the early phases of the transition must evolve if they intend to be effective in the later stages. The Forest Service—one of the world's preeminent land-management organizations—has precisely that opportunity.

REFERENCES

Alverson, W. S., W. Kuhlman, and D. M. Wallen. 1994. *Wild Forests: Conservation Biology and Public Policy*. Washington, DC: Island Press.

Binkley, C. S. 1993. Adapting to Global Supply Constraints: Timber Famine and Six Reasons Why It Won't Occur. In *The Globalization of Wood: Supply, Processes, Products, and Markets.* Madison, WI: Forest Products Society.

———. 1997. Preserving Nature through Intensive Plantation Management. *Forestry Chronicle*, 73: 553–8.

———. 1998. Forestry in a Postmodern World or Just What Was John Muir Doing Running a Sawmill in Yosemite Valley? *Policy Sciences Journal* 31: 133–44.

Cronon, W. 1996. *Uncommon Ground: Rethinking the Human Place in Nature.* New York: W.W. Norton.

Diamond, J. 1997. Location, Location, Location: The First Farmers. *Science* 278: 1243–4.

Farnum, P., R. Timmis, and J. L. Kulp. 1983. The Biotechnology of Forest Yield. *Science* 219: 694–702.

Ghazali, B. H., and M. Simula. 1996. Study on the Development in the Formulation and Implementation of Certification Schemes for All Internationally Traded Timber and Timber Products. *International Tropical Timber Council Report.* ITTC(XX)/8. Manila, Philippines.

Heun, Manfred, Ralf Schäfer-Pregl, and Dieter Klawan. 1997. Site of Einkorn Wheat Domestication Identified by DNA Fingerprinting. *Science* 278:1312–14.

Hyde, W. F., D. H. Newman, and B. J. Seldon. 1992. *The Economic Benefits of Forestry Research.* Ames: Iowa State University Press.

Kellert, S. R. 1993. Attitudes, Knowledge, and Behavior Toward Wildlife Among the Industrial Superpowers: United States, Japan, and Germany. *Journal of Social Issues* 42.

———. 1995. Concepts of Nature East and West. In *Reinventing Nature: Responses to Post-Modern Deconstruction*, edited by M. Soulé and G. Lease. Washington, DC: Island Press, Chapter 7.

Kimmins, J. P. 1994. Ecology, Environmentalism and Green Religion. *Forestry Chronicle* 65: 285–9.

Lyon, K. S. 1981. Mining of the Forest and the Time Path of the Price of Timber. *Journal of Environmental Economic Management* 8: 330–44.

Neilson, D. 1997. *The Tree Farm and Managed Forest Industry.* Rotorua, New Zealand: DANA Publishing.

Perlin, J. 1991. *A Forest Journey: The Role of Wood in the Development of Civilization.* Cambridge, MA: Harvard University Press.

Proctor, J. D. 1996. Whose Nature? The Contested Moral Terrain of Ancient Forests. In *Uncommon Ground: Rethinking the Human Place in Nature*, edited by W. Cronon. New York: W.W. Norton.

Pyne, Stephen J. 1997. *World Fire: The Culture of Fire on Earth.* Seattle: University of Washington Press, Henry Holt & Co.

Sedjo, R. A., and D. Botkin. 1997. Using Forest Plantation to Spare Natural Forests. *Environment* 39(10): 14–20, 30.

Discussion

Forestry in the New Millennium

Kenneth L. Rosenbaum

In his discussion of park management, Clark S. Binkley notes three categories of problems: philosophic, managerial, and technologic. I offer a set of more general categories that accommodate, with a little stuffing and shoving, the broader set of issues of forest management that he discussed.

- Technical issues: What do we know? Do we know enough to manage the forest? (This category includes Binkley's managerial and technologic issues.)
- Value issues: What do people want? (This category includes Binkley's philosophic issues.)
- Choice issues: Given competing goals, how do we decide? (This category includes a whole set of institutional issues that Binkley discusses throughout Chapter 5.)

WHAT DO WE KNOW?

On the technical issues, Binkley and I more or less agree. Whether society's aim is timber or biodiversity, we do not know all we want to know about forest management. If we knew more, we could produce more. Perhaps we could avoid some of the conflicts among uses, too, but new knowledge also can lead to new conflicts.

Binkley makes some excellent points about investment in research. The private sector underinvests in general forest research, and govern-

---

KENNETH L. ROSENBAUM is principal consultant with Sylvan Environmental Consultants, Washington, DC.

ment must take up the slack. Also, the government must pursue those research topics that do not interest the private sector but are crucial to public lands management. However, knowledge can grow from activities other than formal research. The Forest Service is beginning to formally apply adaptive management. Adaptive management involves framing management actions as carefully monitored experiments designed to inform and improve future management decisions (Lee 1993). The Forest Service should do more adaptive management.

Adaptive management is hard to do. When you embark on adaptive management, you have to admit that you do not know all there is to know and that you may make mistakes along the way. You have to be willing to invest in monitoring and evaluation that may prove you to be completely wrong in your actions, and you have to be prepared to admit it and start again. The Forest Service has a less than perfect track record on budgeting for monitoring and evaluation (U.S. GAO 1997, 41).

Moving the agency to adaptive management will require strong leadership.

Embracing adaptive management will be a risk in itself. We do not have all that much experience with it. Little beyond its appeal to common sense shows that adaptive management can improve resource management. If we track the wrong indicators or test poorly designed options, adaptive management could even be misleading. But the alternative—to forgo learning all we can from our mistakes—is far more risky.

WHAT DO PEOPLE WANT?

The reader of Binkley's chapter might get the impression that timber is the most valued output of the national forests; that timber is a rather straightforward management objective; and that timber's complement, "nature," defies easy analysis.

When you look at it closely, timber is quite a complex set of outputs that generate conflicts of their own. Management of a stand to maximize wood chip production is incompatible with maximizing saw timber output. Furthermore, it is not obvious what sort of management mix is most desirable. Even the market is not a sure guide. One reason is the temporal momentum of forests.

Temporal momentum, Binkley explains, means that forests have a long production cycle relative to their annual usage. A systems engineer might call temporal momentum a built-in feedback delay. Feedback delays cause systems to oscillate and to behave other than how we want them to behave. So, for example, in the economy at large, the momentum of capital investment often puts production capacity out of step with demand and causes economies to oscillate between expansion and recession.

A forester managing in response to today's demands produces a forest that must meet demands decades in the future. If the future wants what we want, and the management is technically sound, then we are in good shape. But history shows us that the future is hard to predict.

Consider the demand for market goods from trees since the United States declared its independence. In the 1700s, the eastern forests were strategic assets. Long poles from pine trees, naval stores, and oak planks and joists were essential to maintaining a navy. In the mid-1800s, the new necessity was wood suitable for railroad ties and fuel. After the Civil War, New England's woods were a hotbed of charcoal manufacture for industrial fuel. During the second half of the 1800s, we figured out how to make paper from trees, opening a whole new area of demand.

In the 1900s, we have experienced similar changes in demand. Fuel has become a far less important use of wood. New technologies in product manufacture have changed the way we look at the forest. In the early part of the century in the Pacific Northwest, western hemlock was considered a weed tree (Greeley 1951, 123). In the 1950s and 1960s, big-leaf maple, red alder, and the Pacific yew were considered weeds, too. Now, all these former nuisances have found valuable uses.

If you consider nontimber demand, the situation becomes far more complex. In 1900, who managed the national forests in anticipation of demand for skiing and hiking areas? Who foresaw the modern value placed on protection of scenery or biodiversity? Today, even private landowners must comply with public interest regulations.

What will drive future demand? Will the value of forests as carbon sinks (particularly in countries with governments stable enough to reliably promise to maintain lands in forest) influence forest management? Will biodiversity prospectors, looking for new antibiotics and pesticides, trigger a gold rush for the genomes of undisturbed forest soil bacteria?

Will income from underplantings of gourmet fungi, medicinal herbs, and the like come to rival income from the trees themselves? (Pilz and others [1999, Table 2] showed the present net worth of matsuke mushroom harvests rivaling timber on a hypothetical stand in the Winema National Forest in Oregon.) Will new recreational technologies (say, off-trail in-line motorized skates) bring millions out to enjoy the forest?

American foresters always have made sacrifices to meet the needs of the future. At a minimum, we all commit to managing the lands to ensure their future productivity. The Forest Service always has gone a little farther than most private land managers. The national forests aim for a nondeclining, even flow of timber to prevent forest-dependent communities from experiencing future economic busts due to lack of trees of harvestable age. Arguably, the many steps that the Forest Service takes now to conserve biodiversity on federal lands are actions to meet future demands. The whole concept of sustainable development entails present constraints on actions to serve future demands.

But if you cannot know exactly what the future wants, what are you trying to sustain with your management? There are two polar responses to this question. One is that we assume that the people of tomorrow are going to want what we want today. Although we know that this scenario is false, it may be closer to the truth than some other wild-eyed guesses that we might make about the future. We commit to managing for sustainable production of today's forest uses. We figure that our successors will make do with whatever we give them, just as we make do with what our predecessors left us. The second response is a strong application of the precautionary principle: Do nothing today that might constrain tomorrow's choices.

The practical response is somewhere in the middle. But where?

A twist in the problem is that what we save today will influence demand tomorrow. Having lots of merchantable timber on the market will encourage investment in mills and thus promote demand for the timber. Having lots of wilderness available will increase the number of people experiencing wilderness and teach people to use and enjoy it. There is no particular drumbeat today to restore the dammed valley of Hetch Hetchy, but the supporters of protecting the scenery in its twin, Yosemite, are legion. (For thoughts on making environmental decisions today to deliberately shape future environmental values because those values themselves have worth, see Sagoff 1982, 298–301).

This discussion began by looking at timber, but nonmarket commodities somehow crept in. It is inevitable. No forest management situation deals only with market-priced commodities. Even on a plantation, the manager is faced with balancing unfettered timber harvest against protection of soil, water, and other public goods.

Binkley notes that our aims for some nonmarket commodities are ill-defined. He points to the example of reserve management and discusses the philosophic problems in managing a reserve to be "natural."

In my experience, defining what is natural is something of a side issue to most park managers. Like the manager of a timber plantation, the manager of a park or wilderness area needs a practical goal. Out of practicality, management objectives usually are framed in other terms. The manager of Yellowstone Park tries to conserve "vert and venison" consistent with an educational and enjoyable experience for a high volume of visitors. Only when a wildfire or other extraordinary event brings the management debate to a wider circle than just the day-to-day managers do we bog down in a discussion of what natural is.

Even then, on close inspection, the question of what is natural proves to be a red herring. (For more on the false trail of "natural" management, see Budiansky 1995.) The real issue is not debating the philosophic meaning of "natural" but deciding which of the many competing individual values for forest use should be the social choice. This decision involves examining how to make social decisions, which is discussed later.

HOW DO WE DECIDE?

Now we come to the meat of the issue. Many high-profile problems of the Forest Service today come down to resolving conflicts over competing uses. We have competing ideas in this country about how public lands ought to be managed. How do we decide among them?

The method of "deciding how we decide" can take many forms. We can influence decisions by shaping the institutions that make them, by assigning land rights to different owners, or by writing rules of procedure for government. We also can codify some of our basic decisions as laws that apply to public or private forests.

Binkley's central conclusions and recommendations go to the heart of how we choose among competing uses:

1. Natural forest management is contentious and will inevitably decline in importance as a source of industrial timber. We should encourage this decline and phase out natural forest management.
2. The private sector does better than the public sector in timber management. We should transfer prime timber lands to private ownership.
3. Parks and preserves present a major opportunity for the Forest Service, if the conflicting visions for management of these lands can be untangled. The Forest Service should articulate clear sets of goals for noncommodity management of remaining lands and learn to manage the national forests to meet this set of goals.

He is especially critical of "natural forest management." His criticisms deserve a closer look.

Mixing market and nonmarket outputs in natural forest management creates objectives that are incompatible, claims Binkley. But as argued above, even market objectives can be mutually incompatible. Making decisions when objectives conflict is tough, but deciding among incompatible objectives (for example, guns versus butter, tax cuts versus social services) is something governments do every day. That's why we have governments. The real need is to improve how we make those decisions.

Binkley criticizes forest practice codes, saying that they are costly and do not guarantee ecological success. But he mentions only the cost of the code to the landowner, without discussing the social cost of not having the code. He offers one example of an ecologically flawed code, but not all codes are deeply flawed, and even those that are can be amended and improved.

Next, Binkley discounts forest certification as too constraining of harvest volumes. He says that if all forests were certified, then they could not meet current world demand for timber. If this is true, then we need to ask why. Are current certification standards off base? Or are current harvest practices unsustainable?

Natural forest management, Binkley predicts, will lead to progressively lower public revenues, lower Forest Service budgets, and eventually less responsible stewardship. Although user fees and self-supporting programs are increasingly popular in government, most agency budgets are not tied to revenue. For example, police departments are not required to support their budgets with the proceeds of fines or user fees from crime victims. Similarly, the Forest Service generates a host of outputs that produce no timber or revenues. Agencies that produce public goods should have access to public funding. If the political demand for nontimber forest services is high enough, the Forest Service will have the budget it needs.

Binkley advises moving productive timberlands to the private sector. This shift might reduce conflicts over forest land management in the long term, but we would have a royal battle over which lands to move at the outset. It also would greatly complicate future efforts to manage the forests as ecosystems. No matter what boundaries we choose, nature eventually poses a management problem that requires coordination across those boundaries. The more we divide ownership of the forests, the more difficulties we pose for coordinating their management.

Additionally, Binkley suggests developing management technologies for parklands that better reflect the various social demands and desires for its use. He does not give a great deal of guidance on how to find those technologies, but that is understandable; no one quite knows how. He has identified a genuine challenge.

A few other challenges that relate to forest institutions Binkley does not directly address. The first set of challenges, which might be called macroinstitutional challenges, relate to how the government makes decisions among competing uses. The first issue involves the relative roles of the Forest Service and the U.S. Congress. We currently have some general directives from Congress on management, which the agency tries to implement through a complex and lengthy planning process, and which Congress then revises through the appropriations process. A major challenge for the Forest Service and Congress is to devise a planning system that wastes less effort on internal friction.

Another macroinstitutional issue that we should face is the organization of the current array of public land agencies. If timber production drops low in the rank of agency functions, why keep the land management arm of the Forest Service in the Department of Agriculture? Why have separate agencies managing parks, refuges, and forests? Is it time to reinvent the public land agencies?

Changes in management direction also raise microinstitutional issues. Agencies, like forests, have temporal momentum. When Pinchot created the Forest Service, he wisely brought in all new people, eschewing the old Land Office employees. A recent survey shows striking differences in environmental values between traditional forest managers and noncommodity managers in the Forest Service (Brown and Harris 1998). Can we really have a modern Forest Service—one with much less emphasis on market commodities—without a major, potentially painful change in personnel?

CLOSING THOUGHT

To come full circle, let us end where Binkley began: the meaning of "forest." Johnson's 1701 definition may be the oldest one found in a dictionary, but Dr. Johnson took it almost word-for-word from a legal treatise written more than 100 years before (Manwood 1598, chapter 1, part 1). "Forest" came into the language as a legal term to describe property.

Manwood was careful to explain that a forest entailed more than vert and venison. It included a set of laws and a bureaucracy of officers and courts to enforce them. So, government forests and forest bureaucracies are nothing new. Neither is conflict over forest use. Manwood's book covers laws developed during 500 years of litigation.

The documented histories of conflict and changing demands on the forest are older than the English language. The challenge to the Forest Service is not to eliminate conflict and change, but to thrive despite them.

REFERENCES

Brown, G., and C. Harris. 1998. Professional Foresters and the Land Ethic, Revisited. *Journal of Forestry* 96(1): 4–12.

Budiansky, S. 1995. *Nature's Keepers: The New Science of Nature Management*. New York: Free Press.

Greeley, W. B. 1951. *Forests and Men*. Garden City, NJ: Doubleday and Co.

Lee, K. N. 1993. *Compass and Gyroscope: Integrating Science and Politics for the Environment*. Washington, DC: Island Press.

Manwood, J. 1598. *A Treatise and Discourse of the Lawes of the Forrest*. Facsimile edition, reprinted 1978. New York: Garland.

Pilz, D., and others. 1999. Mushrooms and Timber: Managing Commercial Harvesting in the Oregon Cascades. *Journal of Forestry* 97(3): 4–11.

Sagoff, M. 1982. We Have Met the Enemy and He Is Us, or Conflict and Contradiction in Environmental Law. *Environmental Law* 12: 283–315.

U.S. GAO (General Accounting Office). 1997. *Forest Service Decision-Making: A Framework for Improving Performance*. GAO/RCED-97-17. Washington, DC: U.S. General Accounting Office.

6

State Trust Lands Management

A Promising New Application for the Forest Service?

Sally K. Fairfax

> [A]t the heart of the nation's public land policy one finds a conceptual and operational void. It has existed for at least three generations ... nearly all of contemporary discussion of the lands seems stagnant, unable to move beyond ideas that were already clichés by World War II.
>
> —Frank Popper (1988)

This chapter was written on the premise that one reason for the vacuous nature of public and professional debate regarding public resources is that we have so few words, ideas, and visions for discussing them. Accordingly, this paper will do two things. I first identify the kinds of ideas that will be of most value in reforming national forest management. A brief administrative history of federal resource management is included, with an eye to identifying the forces of change to which future managers of public resources will have to respond. Second, I describe the state trust lands (in general and with some specific examples), emphasizing their fit with the forces of change and what they tell us about possible approaches to national forest management. I believe that the trust model has much to teach that is responsive to current pressure for change.

Although our tools for thinking about public resources have not changed much in a century, our management of those resources has evolved considerably, particularly in the past thirty years. The whole

SALLY K. FAIRFAX is a professor of public policy in the College of Natural Resources at the University of California, Berkeley.

bouquet of outputs from national forest management has been dramatically rebalanced—the emphasis shifted from timber, range, and minerals to increasingly motorized and commercialized recreation and preservation (for example, *High Country News* 1998). Whereas there is no doubt that we are moving in that direction, our vocabulary for discussing public resources remains mired in concepts that were, as Popper (1988) noted, already outdated and irrelevant when I was a child.

This Resources for the Future (RFF) effort to encourage more fruitful approaches to the national forests seems timely for several reasons. First, we are living in an era of Forest Service centennials: In 1997, we enjoyed—with refreshingly little fanfare from the Forest Service—the 100th anniversary of the 1897 putative Forest Service Organic Act, and in 2005, the agency itself will celebrate the 100th anniversary of its founding. Second, and more important, widespread consensus is that the Forest Service is quite simply falling of its own weight. Fifty years ago, the agency was generally regarded as one of Uncle Sam's resounding successes. (For a little perspective on how far the mighty are fallen, see a précis of a gushy five-page *Newsweek* article on the agency from its June 2, 1952, issue [Forest Options Group 1998].) Today, it is difficult to find many people—in or out of the agency—who are willing to feebly protest the suggestion that the agency will not make it to its 100th birthday or, sadly, even to find more than a few people who think it matters much. The Forest Service's reputation is marred by scandal, and Forest Service "abuse is a favorite sport on Capitol Hill.... Even Smokey the Bear [sic] is blamed for many forest health problems."[1] No bangs and little audible whimpering accompany the agency's lackluster deflation.

It is not clear to me that windy pontifications from academics—even from so keen an observer as myself—are called for to rectify the present situation. However, I am delighted to participate in this effort to help move the conversation off dead center by discussing the state trust lands. For the past fifteen years, a series of graduate students and I have been exploring state management of public lands granted to them as part of each accession bargain. The grant program lasted from 1803 (Ohio) until 1959 (Alaska). The general pattern (to many aspects of which, Alaska, Hawaii, and California are exceptions) was an explicit bargain: In return for waiving all "right, title, and interest" in the public domain lands within its boundaries and agreeing not to tax recently patented lands, the U.S. Congress granted to the states, among other things, one to four sections in every township for the purpose of supporting public schools. Additional lands were granted to support hospitals, prisons, insane asylums, and similar public institutions, depending on the deal that the state was able to cut at the time it joined the Union. (For more detail on the history and current management of these lands, see Souder and Fairfax 1996.)

State trust lands constitute a major, if generally unseen, land management regime in the United States. Twenty-two western states manage approximately 135 million acres—closer to 155 million if you include lands in which the states hold only the mineral estate under the trust mandate. Compared with the holdings of the federal agencies (National Park Service, 80 million acres; Fish and Wildlife Service, 100 million acres; Forest Service, 192 million acres; Bureau of Land Management [BLM], 270 million acres), the state trust lands emerge as a significant—and by more than 100 years the oldest—public land holding and management system in this country.

Moreover, the management mandate for those lands is strikingly different from that afflicting the Forest Service. These state lands are held in trust; not the public trust, à la Joe Sax (Sax 1970b; I am told that this article is one of the ten most cited law review articles in history), but a Merrill Lynch–type trust. A trustee manages resources for the exclusive benefit of a designated beneficiary under strict rules of disclosure and accountability enforceable in the courts. This mechanism is approximately the same as when a grandparent designates funds to be expended for an heir's college education. The goal is clear, the disclosure requirements are detailed, and the process for accountability is familiar and, in the presence of an aroused beneficiary, effective. Crucially, when comparing trust management and federal agency programs, there is no deference to the trustee's alleged expertise in this judicial oversight. Trustees must, at a minimum, evince the prudence of an ordinary person. A trustee who possesses or claims to possess any superior talents or abilities only inspires the courts to hold the trustee to a higher standard of care (discussed in Souder and Fairfax 1997).

The trust lands model meets my criteria for a new idea despite the fact that it is our oldest land policy. Indeed, the trust is being bandied about in many settings and permutations as a nostrum to cure at least some of what ails public resource management (for example, Forest Options Group 1998; Hess 1993) and is applied in numerous and interesting settings. Whereas trusts operate on state lands and are occasionally discussed as an alternative on federal lands, trust principles are being applied in diverse contexts. My recent work focuses on the use of trusts and trust principles specifically to accomplish conservation goals. These range from the *Exxon Valdez* Oil Spill Trustee Council (EVOS), an organization established to oversee expenditure of the nearly $1 billion paid in damages by Exxon to the federal government and the state of Alaska, to less well endowed but equally interesting family trusts established to preserve particularly cherished parcels of land or similar resources. Trusts are being discussed and used in the context of germ plasm conservation, a plethora of project mitigation settings, and the protection of indigenous peoples. The flexibility of the trust instrument is impressive and important (Guenzler and Fairfax forthcoming).

However, if I have learned anything in the past twenty-five years, it is that scholars with new and improved theories about policy are probably overpriced at a dime a dozen. Any new thoughts about national forest or public land management will enter a policy arena with all the trump holders fairly well paid off.[2] A new idea must either capitalize on shifts in public mood to reflect a new and applicable consensus or be able to generate a new constituency that significantly alters the existing balance of power. Accordingly, I begin by discussing three major forces that could alter the current balance of power sufficiently to constitute an "increment" in national forest management.

All three forces reflect institutional change in the public resources field. The first is that the courts, which have dominated the public lands policy arena since the late 1960s, appear to be exiting the arena. The second force is a decline in federal agency efficacy, which has two elements. The Forest Service is suffering from the loss of buoyancy in both compartments of its Mae West: Science[3] and the federal government are eroding as sources of authority, and with them goes much of the infrastructure of federal legitimacy. The third force, very much related, is the institutional fragmentation in an operation that has been, for most of the century, dominated by the Forest Service. Government institutions in the conservation field have fragmented, creating enormous openings for states and localities. More interesting, perhaps, is the growing irrelevance of the national environmental groups that have shaped debate since World War II. To the extent that the national environmental groups are an amalgam of preservation and recreation interests, they probably are doomed. Recreation is no longer—if it ever was (Joe Sax notwithstanding)[4]—an aesthetic, simplifying undertaking. It is moving rapidly toward a mechanized, industrialized enterprise that has little to share with preservationists. It never has been clear to me that this alliance was a marriage of true interests—I always thought of it in terms of a successful kidnapping—but I believe it is over. Community control and economic efficiency (odd bedfellows) are part of an effort to fill a void created by the multifaceted decline of the federal government—courts and agencies—as actors in the public resources field.

After highlighting these three drivers of change in a selective and abbreviated historical review of national forest administration, I will discuss the state trusts as a response to those forces. The trust is interesting in the context I have described for several reasons. First, judicial oversight of trustees and the fulfillment of the trustees' obligations are on a significantly different track from the court's review of agency discretion. Second, the trust has generally relied on the market rather than the world of science as a metaphor for what the trustee ought to be doing. Trust principles require an honest risk–benefit assessment instead of a series of term papers averring a certainty in analysis and outcomes that is not credible. They also rely on issues of profit and loss to define accountability. This businesslike element of the trust is what gives it the most appeal to new-right economists and the Republicans in Congress. They find it tempting, I fear, to substitute their rather naive catechism about market forces for the Pinchot–Roosevelt ideal of the perfect scientific government. Beyond allowing us to examine different metaphors, the trust can teach us rather quickly that these tools do not translate easily into improved management. Third, and most interesting given my emphasis on institutional fragmentation, the trust is instructive because of its flexibility. Trustees are not lashed to a mast of one-size-fits-all policies regarding pricing, access, leasing, or even public involvement. The opportunity to observe different concepts and tools in operation in an altered institutional setting is important. It is in the opportunity to find small innovations and incremental changes in activity and attitude that I find the trust most valuable as a teacher.

DRIVERS OF CHANGE

I have divided the twentieth century into seven periods, each of which reflects important elements in the shifting institutional setting of federal agencies.

National Government and the Administrative State

During the final third of the nineteenth century, the federal government began to grow, relative to indicators of geographic and population explosion, in both size and scope. Not surprisingly, the growth in size was related to a growth in stature. Confronted with (1) the emergence of a national economy and national corporations and (2) the need to address the social disarray caused by industrialization and rapid urbanization, the federal level of government displaced a political culture tied to localities and states and emerged as the focus of political action. It began doing things that government at any level simply had not done before (see Wiebe 1967).[5]

A major focus of this reorientation was public domain policy. As Joseph Sax pointed out, until the 1890s, it was a serious question of constitutional law whether the federal government could acquire land for purposes of Civil War memorials (Sax 1984, 126; see also Connick and Fairfax draft). It also was not until the late nineteenth century that the

U.S. Supreme Court held—and rather ambiguously at that—that the federal government was authorized to make regulations regarding the use of federal public land in the western states and territories.[6] A key element of the growing faith in the federal government was a simultaneous emergence of science as a basis for legitimacy in public affairs. The discussion of these changes in Samuel P. Hays' *Conservation and the Gospel of Efficiency* (1959) focuses on the progressive era's embrace of science and "organized, technical, and centrally planned" government activities.

Nondelegation, and Why the Forest Service Prospered Anyway

Students of the conservation movement are misled if they follow Hays farther than he intended. It is important to underscore that the Progressives were generally not successful in converting the Supreme Court to progressivism (Wherhan 1992). In the first third of this century, the courts fairly consistently rejected agency programs and decisions by locating in them an unconstitutional delegation of authority.[7] The U.S. Forest Service successfully dodged the nondelegation bullet. Public domain management was one area in which the court significantly aided the expansion of federal authority. The court oversaw a rather rapid expansion of the Article IV property clause of the U.S. Constitution to permit the federal government to acquire property within states (as opposed to within the District of Columbia or territories) and to retain, or decline to dispose of, public domain lands.

The question of why the Forest Service emerged an early winner in the nondelegation terrain is interesting. It seems important that the agency presented itself as the premier expression of scientific decision-making. It is interesting, for example, to compare the agency's scientific emphasis with the court's apparent concern for the lack of data supporting the regulations challenged in cases where federal regulation was disallowed.[8] This reliance on science is the core of a fundamental fiction that evolved during the first decades of the twentieth century: It was possible in our democratic society to allow executive agencies a large role in American life and the American economy because they did not enter the political arena. The Forest Service benefited from that idea.

Administrator Changes from Expert to Voice of the Public Interest

The court's disapproval of administrative agencies and the delegation of legislative authority continued. Early New Deal programs were disallowed until President Franklin D. Roosevelt aired his famous "court packing" scheme.[9] Almost immediately, the Supreme Court, and the judiciary branch more generally, began looking less carefully at delegations of congressional authority to executive agencies. This change in judicial position was justified by two theories that reflected a growing acceptance of federal administrators: (1) a presumption that the agencies represented both expertise *and* the public interest and (2) judicial deference to agency expertise in defining broad programs to address broad social needs. The Forest Service parlayed its technical forestry expertise into an enormous role in economic relief during the Great Depression.[10]

The Administrative Procedures Act: Process-Based Review

In the late 1940s, much of the depression-era thinking about the role of public agencies was formalized into the Administrative Procedures Act (APA). That statute provides a formal basis for judicial review of agency action. The APA tells agencies how to write rules for describing programs and defines the relationship between executive agencies and reviewing courts. All of the familiar process—rules about notice-and-comment rule-making, maintaining an adequate record of factors assessed during the decisionmaking process, and the familiar standards of judicial review of agency discretion ("arbitrary and capricious")—were defined in that statute and its early elaboration.[11] Courts under this model "defer" to "agency expertise." And agency expertise, as was demonstrated in early environmental cases such as the classic *Scenic Hudson Preservation Council v. FPC* (1965), was manifest in an adequate record of decision. At first, the APA did not apply to land managers—in fact, they were specifically exempted from its provisions.

This immunity was probably not important given the few Forest Service cases that found their way into the courts. However, as is familiar, the Forest Service was beginning to cut timber seriously for the first time in its history, and the 1950s also saw the emergence of the wilderness movement and the introduction of the first wilderness bill. Although preservation advocates no longer trusted the agency to protect wilderness areas set aside during the 1920s and 1930s, in the 1950s they rose to defend the agency in the protracted and spectacular early Sagebrush Rebellion—the Barrett–McCarran Hearings (Peffer 1951; De Voto 1955).

The Interest Representation Model

Science emerged tarnished from the *Silent Spring/Structure of Scientific Revolution* era of the 1960s and 1970s. Experts were no longer trusted to speak for the public interest (Horwit 1994, 148), and the administrative arena emerged "as a forum for interest representation, where the public interest would be arrived at through a decisionmaking process to which all relevant groups had appropriate access" (Horwit 1994, 144; see also

Stewart 1975). The notion of "standing" as a constraint on who could bring a suit against an agency all but disappeared during the 1960s and 1970s; where it did not, Congress wrote specific provisions that allowed citizen challenges and, occasionally, citizen enforcement of agency mandates. The court also adopted a rather expansive view of its own capabilities and role, sometimes known as the "hard look" doctrine; courts no longer deferred to agency expertise but quite willingly applied their own notions of reasonableness to create standards for evaluating agency behavior.

These new standards affected the public lands dramatically. No longer exempt as "proprietors," the Forest Service was constantly in court. The Ford Foundation led in funding a breed of National Association for the Advancement of Colored People (NAACP)-like environmental organizations focused on litigation as a reform strategy. Public attention was expanded by a series of successful efforts to halt projects by imposing a newly elaborated form of the APA's record of decision requirements as codified in the National Environmental Policy Act (NEPA). (For a discussion of the relationship between NEPA and the APA, see Fairfax 1978. Joseph Sax [1970a] wrote the major text on using the courts in *Defending the Environment.*) In addition, protest strategies pioneered in the civil rights movement were adopted by post–Earth Day environmental groups, and the clearcutting controversy forever altered the Forest Service's public image and political setting.

New Judicial and Scientific Humility

We are experiencing a reformation on numerous fronts. The most familiar is standing; in several cases that are complex and difficult to interpret, the basic notions that erased standing as a barrier in the early days of the "environmental era" are being recast. This means that fewer environmentally oriented litigants are finding their way into court.[12] Those that do pass muster will find courts less and less likely to overturn agency action. The courts are abandoning the "hard look" doctrine and returning to the posture of deferring to agency decisions. The issue here is not one of enhanced deference to agency expertise; no ennobling or empowering of the agencies is involved.

Instead, the issue is one of separation of power. Basic constitutional questions, long parked at the side of the administrative law road, now are returning to the forefront of discussion.[13] The Supreme Court concluded that the judiciary branch's role was restricted to enforcing the "unambiguously expressed intent of Congress." If there is ambiguity, then the issue is not of law but of policy. And it is the role of Congress to write legislation, and the role of administrators and the executive to interpret policy (Wherhan 1992, 840–5). Ironically, having been one of the early successes in gaining court approval at the start of the century, the Forest Service litigation has been slow to experience the ebb of judicial scrutiny.

Loss of Federal Agency Legitimacy

One would think that this contraction of the court's oversight of agency decisionmaking was potentially good news for the Forest Service; finally, we can get "forestry out of the courtroom and back into the woods." That is a possible outcome, but not a likely one. Unlike in the New Deal era, the current humility of the courts is specifically not accompanied by an embellishment of the agency expertise or authority. Quite the opposite, it is accompanied by two corrosive factors: humility in the ecology profession and a related decline of respect for the federal level of government.

Growing doubts about the political potential of scientific findings originate less with members of a skeptical public—who appear, reasonably, to embrace scientific findings that comport with their biases and to reject those that do not—than within the scientific community itself. The fly in the ointment is, of course, the ecology profession's rejection of long-revered models that are based on concepts of equilibrium in natural systems. (Two excellent sources on "new ecology" as it affects the law are Meyer 1994 and Rodgers 1994. For a deeper background, see Worster 1977.) The new *dis*equilibrium approach translates into ecologists who are (1) becoming less and less willing to attribute specific events or outcomes to particular inputs or to predict the results of alternative strategies on particular systems in relevant time scales and (2) appearing less and less credible when they try to do so. (See Nelson 1995 for more information on the decline of the progressive era and *Greenwire* 1998a for a brief insight into why science is revered but no longer useful in resolving disputes.) One outcome is that the Forest Service can no longer turn to science to trump its critics or define credible technical solutions to complex social problems.

The legitimacy of federal agencies is further undercut by changing thoughts about the federal level of government in general. Part of the reconsideration of the virtues of federal action arises because the Forest Service is running out of money to spend and/or encountering a growing unwillingness to spend it. (My favorite source on this phenomenon is the only slightly outdated O'Toole 1994.) The federal lands are eroding a reliable conduit of subsidies to states and localities (Fairfax 1987).[14] This erosion has enormous implications for the acceptability of the federal presence in the western states (see, for example, *Greenwire* 1998b).

This uncharacteristic emphasis on money is an element related to the federal decline: the diversification of government policy instruments and a corresponding diversification of private organizations at work in the area. The federal level is no longer viewed as the sole or even necessarily the primary agent of public resource management.[15] It is surrounded by rivals at the state and local levels, forced to deal with an increasing array of public and private "partners" and challenged by increasingly legitimate "community" groups seeking not simply to be heard in federally defined planning arenas but also to share and exercise authority.

The erosion of federal hegemony is familiar, but it is accompanied by many related institutional changes in the conservation field (Fairfax and others 1999). *Devolution* of federal authority over federal resources is proceeding at a pace that would likely surprise those familiar with the specific rejection of any transfer of title proposals to come before Congress. Minerals management provides an interesting perspective on the growing state role in federal land management. Devolution does not, in my lexicon, include the dual regulation that has long been an important element of public land management (see Cowart and Fairfax 1988).[16] It refers to areas where the states or localities are exercising authority formerly exercised by the federal agencies and transferred, more or less officially, to other levels of government.

After almost thirty years of dispute, in the mid-1980s, the states began to take responsibility for oil royalty accounting on federal lands. Following the report of the Linowes Commission in 1982 (U.S. Commission 1982), Congress designed a system under which interested states could assume "primacy" in royalty accounting. (The Federal Oil and Gas Royalty Management Act [FOGRMA] of 1982 is discussed in Fairfax and Yale 1987, 73–76.) Conflict on that front continues and has expanded (see, for example, *Public Lands News* 1997b). Several states are now seeking "primacy" in inspection and enforcement in "management of BLM's oil and gas lease program." The BLM position is not to oppose such state participation but to transfer only authority over inspection. The states maintain that BLM "should transfer substantive authority, *much as the Office of Surface Mining has done for coal mining*" (*Public Lands News* 1997a [author's italics]). Much but not all of this authority shifting comes from Congress (see Uram 1998). The federal sovereign is undoubtedly supreme, but the growing extent of state regulation and management of federal resources would surprise most casual observers.

Even more surprising to students of Ted Lowi's 1969 classic, *The End of Liberalism* (in which the author describes and laments the slippage of government authority into private interest group hands), is the extent to which federal programs on federal lands are proceeding in "partnership" with private groups and corporations. Even a casual perusal of a news source such as *Greenwire* over a week or ten-day period will produce at least a handful of new programs in which federal authorities and resources are being "shared" with diverse nongovernmental entities. The proposals of the Quincy Library Group (QLG) are probably the most famous attempt by an outside group to effect control over national forest management. But the QLG's efforts are packaged in legislation and therefore are less stunning as an example of what is occurring on the ground than less publicized partnership examples.

Finally, the institutional fragmentation is apparent in the gradual unhinging of one of the most important alliances in natural resource management. The preservationists have, at least since the 1950s, been able to throw their blanket over a broad array of recreation interests and appear as a solid phalanx of public support for wilderness and, more recently, endangered species as a tool of preservation. This alliance is likely to come unstuck under two kinds of pressure. First, recreation is important money—especially to local economies and agencies that increasingly are looking for user fees. Recreation as a freebie, emphasizing a back-to-nature wanderlust, will evolve into the sine qua non of funding public resource management. Second, emergent recreation emphasizes mechanized/industrial pursuits not compatible with the wilderness. My generation, which put its shoulder behind wilderness legislation and designation issues, is now self-actualizing in sport utility vehicles (SUVs) while younger and more energetic types are seeking access for all manner of all-terrain vehicles, personal watercraft, and the like. At one time, it was relatively easy for us voluntary (and temporary) simplicists to invoke the proverbial factory worker's recreation need to support our much classier preferences.[17] This separation will no longer be possible: A new generation of prosperous young professionals is embracing JetSkis, Humvees, and their functional equivalents. The commodity–wilderness battle of the future will be with mechanized recreationists—not the timber, minerals, and grazing interests.

What Does It Mean for Drivers of Change?

This long passage through history and current events identifies fundamental changes afoot in the public resources field that will define the path to successful reform. What will happen as a result of all these changes in the wind? The incrementalist in me responds, with some relief and some sadness, nothing precipitous. Change during the progressive era came gradually, over a period of fifty years or more and not really succeeding until the New Deal. Thus, I do not believe that the trust principles I am about to discuss ought to be on the table so we can decide, next month or next year, to turn the national forests into trusts. More likely, there will be a little turn on a dozen knobs and dials, and a new set of standard operating procedures and standard operating assumptions will gradually displace the old ones—not entirely, but enough to affect policy.

Some of the dials that I consider more relevant include the following:

- The courts will play a decreasingly important role in Forest Service decisionmaking. Their ability to enforce a "single" national vision that emphasizes environmental values will continue to erode.
- The federal level will recede as *the* source of priorities and subsidies. The Forest Service will be reduced in size, importance, and leverage. Similarly, national environmental groups, whose importance has been tied to the stature of national agencies, will be increasingly challenged by local and regional groups.
- Local priorities and processes will increase in legitimacy and in political impact. This change will mean fewer one-size-fits-all rules and institutions, a more diverse planning and decisionmaking process, and more comprehensive values and management goals.
- Sustaining rural economies will become the goal of public forest management. This concern will no longer function primarily as a cover for transfer payments to the timber industry and will be balanced by a continuing insistence on ecological sustainability.
- Federal lands commodities will be defined less in terms of timber, mining, and grazing and more in terms of economically productive mechanized and industrial recreation. This trend will exacerbate the split in the preservation/recreation coalition and will further devalue national environmental groups.

A SPINNAKER TO CATCH THE WINDS OF CHANGE

These emerging conditions are well-suited to the state land trusts and, more generally, to the trust mechanism. In this section, I describe the trust in as little detail as possible and focus on lessons in three areas of change: a different role for the courts, the emergence of the market as a rival metaphor to science in Forest Service arrangements, and the diversification of institutions as the federal level becomes less important.

State trust land agencies are structured significantly differently from the federal ones. State programs bear a superficial resemblance to federal land management—historically, they emphasized grazing (the most extensive trust lands use), timber, minerals, and recreation; they have ignored water as a trust resource; and, for most of their history, they have been dominated by lessees seeking subsidies off of publicly held resources. However, beginning in the 1950s, lawsuits relatively rapidly reoriented the priorities in most trust management organizations, and developing programs for sustainable support of the beneficiaries is now the major element of trust lands management culture.

The institutions that manage the trust lands vary enormously. Some are headed by an elected commissioner, some commissioners are appointed by the governor, and some are selected by a board of trustees or directors; some are one element of a more general resource management and/or regulatory organization, and some are freestanding trust lands organizations. All gather revenues from trust lands management activities and distribute them, under different rules, to the beneficiaries. As a part of that fund distribution, all states have "permanent school funds" into which trust lands receipts are deposited, but only in two cases does the same organization manage the lands and the funds. Generally, revenues from the sale of land and nonrenewable resources are deposited in the permanent fund and only the interest is distributed, whereas rentals and receipts from sales of renewable resources are distributed annually. It is generally asserted, with only partial accuracy, that the land offices pay expenses with receipts. Some do, but the state legislature generally appropriates some percentage of receipts to be expended by the land office. (For a summary of the precise system in each state, see Souder and Fairfax 1996, 45–47.) In presenting the annual budget request, the trustee is required, among other things, to present expenditures in the context of their contribution to returns; however, the state land trusts activities are not typically funded directly from receipts.

The trust is always an appealing organizational option, but in the context I have described, it has particular relevance. It is responsive to the market, and many of its fundamental elements presume (but do not require) that benefits are going to be described in monetary terms. Nevertheless, the fundamental commitments of the perpetual trust are to (1) the long-term sustainability of the productive capacity of the trust corpus and (2) undivided loyalty to the beneficiary. Most of the trust rules work to ensure that the trustee does not enrich him- or herself or others at the expense of the beneficiary. The trust is flexible and can be adapted to structure almost any arrangement of ownership, management, and access. Furthermore, by carefully thinking about the beneficiary, it is possible to design an interesting variety of incentives among participants. The trust has limits—observers are frequently tempted to present the trust instrument as a trump on politics—but it is more accurately described as a new and different kind of insulation against the problems that seem most to afflict us now.

What Are Trusts, and How Are They Different from Federal Lands?

The trust is a species of a fiduciary relationship. That is, a trustee holds and manages property, under exacting rules, for the exclusive benefit of another. This kind of trust is probably most well-known in the context of a grandparent directing a bank or other guardian to manage specified resources such as a trust fund to assure that the grandchildren have sufficient resources to go to college or to achieve some similarly defined purpose. (Guenzler and Fairfax [forthcoming] distinguish what I have come to call a beneficial trust from other kinds of trusts, such as public trusts, sacred trusts, and land trusts.) The rules for administering a trust are clear, relatively easily summarized, and focused primarily on ensuring three things: (1) that the trustee does not enrich him- or herself or others with trust resources, (2) that the trustee does not fritter them away with excessive management manipulations, and (3) that the trustee does not allow trust resources to lie fallow and go to waste. The rules can be summarized in five categories: clarity, accountability, enforceability, perpetuity, and prudence (Souder, Fairfax, and Ruth, 1994; Souder and Fairfax 1996).

Clarity. The purpose of the trust must be clearly stated and include the fundamental element of "undivided loyalty." The trustee is obligated to use and manage trust resources to achieve trust purposes for the *exclusive* benefit of the designated beneficiary. This rather stark command has genuine appeal compared with the multiple-use mandate that directs both the Forest Service and BLM to manage the resources under their authority "in the combination of uses that best meet the needs of the American people" (16 U.S.C. SS 531-4(a); discussed in Souder and Fairfax 1996, 348, note 74). This ambiguous mandate gives agencies enough discretion to engage in below-cost timber sales, grazing leasing, and recreation programs—subsidizing activities that benefit powerful constituencies and create jobs and enhanced budgets for themselves. Such activities would be carefully scrutinized under the undivided loyalty standard of a trust.

Accountability. A second element of the trust mandate—again, one that clearly distinguishes it from most federal government agencies—is the notion of accountability. The rules for disclosure of accounts are extensive. Trustees must produce data about investments and returns that make it possible for the beneficiary to evaluate the trustee's management of the corpus. Period. The requirements are so strict that they are frequently described as tantamount to a "rebuttable presumption of fraud or undue influence" (discussed in Bogert 1987).

Enforceability. It is clear that these duties are enforced. Trust principles are not merely hortatory expressions of good intentions. Unlike the "whereases" at the beginning of legislation or the lofty aspirations expressed in a memorandum of understanding, the trustees' duties are *obligations*. It is important to remember that classic trust enforceability presumes and depends on a beneficiary who will be vigilant in monitoring the trustee to protect his or her interests in the trust.

Perpetuity. Trusts are not necessarily perpetual. There is little reason to extend the college student's trust beyond his or her matriculation. However, the state trust lands and most of the trusts I have studied under the general heading of "conservation trusts" are perpetual, that is, intended to produce benefits forever. (The normal "rule against perpetuities" does not operate against charitable trusts, discussed with probably illegal brevity in Guenzler and Fairfax [forthcoming].) Thus, the trustee is not allowed to prefer any generation of beneficiaries over any other. The commitment to perpetuity in the context of land management, given the clear obligations of the trustee, comes fairly close to a legally enforceable sustained yield requirement (Souder, Fairfax, and Ruth 1994). Simply to create an environment where the present statutory direction regarding sustained yield is enforced would move national forest management in radically different directions.

Prudence. The courts evaluate trustee behavior against a standard of prudence.[18] It requires trustees to do different things in exercising judgment than does the arbitrary and capricious standard.[19] Rather than turning "explicitly on the volume of data accumulated to support a specific decision, when alternative courses of action are available," the rules of prudence require the trustee to incorporate analysis of risks and benefits (Souder and Fairfax 1997, 180). Modern prudence emphasizes the trust as a portfolio as a tool for achieving risk management.

This summary does not obviate the need to consult an attorney if one is going to establish a trust, but it gives us a place to start in thinking about how the trust land manager's mandate is different from the multiple use mandate that afflicts the Forest Service.

Trusts in the Courts

The Trustees' Relationship with the Reviewing Courts. The relationship between the courts and the Forest Service has been, in general, a polarizing and frustrating one. Therefore, it is important to understand that trustees have a significantly different relationship to the courts than do administrators. The distinction arises principally from two sources.

First, the trust is best understood as an element of private law; it is a special kind of contract enforced in the courts. Thus, when reviewing trustees' actions, the court is not required to consider its judicial role as against those of the more "political" branches, and little attention is paid to issues of appropriate delegation of authority. Because I have argued that standing doctrine is contracting in the administrative arena, it is interesting to note that in the state trust lands, standing is almost never an issue.

Any schoolchild, parent of a schoolchild, or taxpayer has, in most jurisdictions (but not all; for the implications of the restricted notion of standing in Idaho, see Fairfax 1997), achieved standing without any debate. Were trust principles to bleed into federal land management they would, in this context, tend toward a continuation of the status quo ante, perhaps giving potential Forest Service litigants a bit of breathing room in the present ebb in the standing doctrine. This change alone might inspire litigants to push public land litigation into the mode of trust principles. To understand how judicial standards of review for trustee decisions work in practice, it is useful to compare them with the standards that the courts use for evaluating agency behavior under the APA. (This comparison is treated in Souder and Fairfax 1996 [277–8] and in more detail in Souder and Fairfax 1997.) Court review of executive agency decisions is focused on four questions:

- Does the plaintiff have standing to sue?
- Was the action authorized by law?
- Were proper procedures followed?
- Was the agency's decision reasonable?

Although the government loses a disappointing number of cases because there is no statutory basis for the action or because the agency failed to follow the announced procedures, the issue of reasonableness—or arbitrary and capricious, as the APA calls it—ought to be most important.

Administrators accustomed to discussing capriciousness have much to learn from the courts' application of the prudence standard to trustees. First, the trustee bears the burden to demonstrate that he or she has acted *prudently*. The burden of proof regarding proper behavior is on the trustee, not on the plaintiff. Second, evincing prudence is not the same as avoiding arbitrary and capricious behavior. As noted earlier, the prudence standard invites the trustee to explore and understand the risks in a decision and to make judgments designed to minimize them. The trustee is not invited, as the administrator appears required, to collect data to support a feigned certainty that is not justifiable by the facts. The trustee is, however, pushed to suggest how the decisions made minimize risks.

Third, the trustees' expertise is not a shield that requires judicial deference. Quite the opposite, it is an invitation to the courts to demand that the trustee meet a more exacting standard. A trustee is expected to use ordinary skill in managing trust resources, the same as he or she would use in managing his or her own business of character and objectives similar to the trust. However, if the trustee is an expert in a field, or has represented that he or she possesses "unusual capacities," then he or she will "be expected to use them in the performance of the trust" (Bogert 1987, 334–5). Finally, the trustees do not enjoy deference on their reading of the trust documents. Courts are extremely familiar with trusts, the duties of the trustees, and the legitimate expectations of the beneficiary. They have no difficulty defining the duties of the trustee without much guidance from him or her.

When thinking about this process as a whole, the review of trustees appears to proceed on three different levels from the review of administrators: Under the trust, (1) standing is not typically an issue, and contracted standing would not likely affect trust litigation; (2) the burden of proof is shifted from the plaintiff to the defendant; and (3) demonstrating prudent decisionmaking requires the trustee to do slightly different and yet considerably more productive things. The goal is not to defend a "preferred alternative" as the right one, as against all others, but to explore the risks and benefits surrounding a decision and to choose a *mix* of policies that are responsive.

What Could the Forest Service Learn from the Courts' Review? Trust principles ultimately may have some direct relevance to national forest management. Martin Shapiro has been suggesting, with considerable reasonableness in my opinion, that the next place to which the courts will turn for guidance in crafting a working relationship with administrative agencies is the trust standard of prudence (Shapiro 1983, 1988; discussed in Souder and Fairfax 1997). It would be analogous to the court's embrace of interest representation as a standard in the 1970s. The judicial withdrawal I have described is well under way, but it is possible that such a prudence trend could develop simultaneously. Indeed, it is possible that the withdrawal that I have discussed could force environmental litigants to explore this path.

If Shapiro is to be right anywhere, the public lands present an excellent opportunity for a gradual shift in emphasis in the court's standards for review. There is no question that the courts frequently discuss the public lands as a form of a trust and a fiduciary relationship. As a broader range of attentive interest groups become familiar with state trust lands management and get shut out on more familiar routes of appeal, they will have an incentive to try new lines of argument that could provide a vector for carrying the vocabulary and the expectations from one field of public land law to another. I would not dismiss that possibility. Indeed, I would entertain it seriously—not because I expect trust review criteria as a whole to supplement or displace the APA (although I would not underestimate the importance of carrying more or less discrete expectations from one field of public land law to another); rather, I would concentrate on what I could learn in the process about trust principles and how they might help me think about federal public land management and reform.

It is useful to note that concepts have rolled across the boundaries from federal to state trust lands management fairly readily in the past. Resource planning is one area in which federal requirements have bled unmistakably into the realm of trust lands management. Recently, the U.S. General Accounting Office (GAO) compared timber management on federal and state trust lands in Washington and Oregon and concluded, completely wrongly in my opinion, that one reason that the states were so much more cost-effective than the federal agencies was because the states were not subject to the wildly elaborate and inefficient planning requirements that afflict the federal land managers (U.S. GAO 1996).

This conclusion is only partly true. Jon Souder and I looked in considerable detail at the planning process in Washington, which we consider to be the best government-run timber management organization in the nation, and concluded that even though the Washington State Department of Natural Resources (DNR) is not under any statutory obligation to engage in elaborate land use planning, it does so anyway (Souder and others 1998). Participants in the forest policy arena have come, over the past thirty years, to expect a certain kind of process and a certain kind of openness in public resource planning. It would not be prudent, the Washington DNR has concluded (correctly, in my opinion), to attempt extensive timber harvest without investing in a similar process. What is different, as I have tried to indicate, is the kind of analysis that the trustee undertakes and the way it is reviewed in court. Nevertheless, the costly and cumbersome planning process has carried over from national forests to state trust lands. This movement suggests to me that it would be prudent for the Forest Service to expect some carryover in the opposite direction. Let me go one step farther: The agency might consider it prudent to encourage some carryover from trust principles to national forest law.

Where would I look for—indeed, seek—that carryover from trust lands to federal lands? First, in the concept of prudence, important thought patterns would be useful to the Forest Service. I refer specifically to the prudent trustee's embrace of risk balancing. It provides a significantly different framework for developing and presenting alternatives in planning and management discourse. Much of what the agency does in this area is defined in statute and regulation, and therefore would change gradually. Nevertheless, a greater emphasis on exercising judgment, presenting risks, and addressing not a single "preferred alternative" but a spectrum of strategies that balance them would, I believe, enhance both the thought processes underlying management and the usability and the credibility of agency plans.

A second area in which the Forest Service should both anticipate and seek a cross-fertilization between trust lands and public lands law is the identification of the kinds of data that are necessary to meet the trustee's obligation regarding full disclosure to the beneficiary—that is, what I have called accountability. A friend recently attended a conference at which Wes Jackson was quoted—by Wendell Berry, I believe—as describing the coming century as "the age of accounting." So be it. The Forest Service has been hiding the ball on cash flows, returns on investment, income forgone, and so on for most of this century. The cat is coming out of the bag now. The agency will be under considerable pressure to improve methods of disclosure—not only for economic indicators but also for ecological ones. If it were my call, I would look hard at state trust lands experience—annual audits, annual budget presentations to the legislature, internal documents and accounting procedures, and challenges to those audits—for guidance. The courts and a whole range of decision-makers are familiar with trust procedures in this area. Rather than make up something out of whole and self-serving cloth, I would adapt practices from the more successful state programs. The state trust lands managers know how to keep books and make them public, even while operating on appropriated funds. I believe it would be important for the Forest Service to take meaningful steps in that direction.[20]

The third area in which the Forest Service should look to court interpretation of trust principles is for guidance in making the sustained yield aspect of the Forest Service mandate operational. Most people I know have long written off the Multiple Use Sustained Yield Act of 1960 and similar statutory and policy calls for sustaining the yield on national forests as unenforceable boilerplate (Jan Laitos [1998] of the University of Denver Law School is particularly insightful on this matter). But in the trust context, the courts have found ways to hold trustees accountable to this clear requirement (for a lengthy but preliminary discussion, see Souder, Fairfax, and Ruth 1994). This requirement typically has a monetary cast to it; the standard trust rendering of sustained yield might not satisfy all ecological health enthusiasts. However, the requirement is far from a nullity and quite adaptable.

The case law on this issue is not extensive in the trust lands context because the issue emerged only recently. However, the response of courts in Idaho and Washington suggests that the tool is a potentially important one. In Idaho, an environmental group seeking to restrain harvest in a watershed was turned back on standing issues, but the framing of the case is instructive.[21] In Washington, timber interests and beneficiaries attempting to force the trustees to harvest more aggressively lost a preliminary but significant battle. The strategy was to demonstrate that conservative harvest regimes on the Loomis State Forest both risked catastrophic fire and were illegally designed to protect the nonbeneficiary lynx. The state argued that in areas of scientific, economic, and political ambiguity, it is prudent to manage conservatively to protect future benefits for future generations. When operating under a mandate to maintain the productive corpus of the trust, the trustee is obligated to maintain a full range of management options by protecting species of unknown but potential value[22] and also to manage conservatively.

The court supported the state's management program and ruled emphatically that the manager of a perpetual resource is not allowed to prefer one generation of beneficiaries to any other. Not to put too fine a point on the importance of sustainability, this is the only case that I am aware of in which a beneficiary challenging a trustee has lost. This line of reasoning ought to be of special interest to the Forest Service as it attempts to move toward a more ecologically based management regime.

The fourth area in which I anticipate—and seek vigorously—some crossover between trust law and public lands law is in the notion of a beneficiary. If I were the Forest Service, I would look for ways to define beneficiaries that would create effective and diverse local support. The agency has done so—indirectly perhaps—in the past, with its payments to county governments. If the Forest Service is going to move away from timber harvest and is looking for a local embrace of more ecologically oriented management, it probably is prudent not to rely on Congress to continue to provide revenue shares in timber harvests that do not exist. Instead, I would give some thought to establishing management trust funds that can be monitored in a way that will give user groups some stake in the protection of future national forest ecosystems. Exploitation of schoolchildren is not wildly popular among opponents of trust lands programs, but the schoolchildren give the operation some appeal. Can a national forest address differing local priorities by agreeing to put some receipts or fees into an endangered species or a land acquisition trust fund? Could the Forest Service take steps to interpret the Knutson–Vandenberg funds as a trust, with trustees who would manage the funds not as a slush pot for the agency but with the legislator's purpose to guide them? (Forest Service Employees 1996). These steps are worth considering.

Trust principles could be transferred more or less in gross by courts and litigators seeking a new model for review of administrative decisions. The better point, however, is that it is reasonable to anticipate and even to seek increased blending of APA and trust review standards. The four areas where review of the trustee could usefully expand Forest Service thinking (that is, prudence, accountability, sustained yield, and beneficiaries) could provide important new ways for approaching basic management issues.

Responding to the Collapse of Science and the Rise of the Economics Metaphor

The progressive era model, as elaborately discussed earlier, provided the Forest Service with a model for organizing its relations with the outside world. The agency's legitimacy began with its embrace of science and has suffered enormously from its eroding credibility. The trust has built less on scientific models than on economic ones. Most obviously, unlike the Forest Service, the very purpose of trust lands management is to make a profit to be used in running "common schools" within the states. I have argued, frequently in concert with Jon Souder, my coauthor on most trust matters, that this market orientation itself is sufficient to transform many of the agency's current ills. We have not gone so far as to argue that it is also necessary, but others have. It has become commonplace in most contemporary discussions of public land reform to embrace market mechanisms, fees for services, full-cost pricing, proper incentives, and so forth. I have little to add. Thus, although one does not want to ignore the role of economic incentives, how they operate in the state trust lands, and what they could teach the Forest Service, it would be silly at this stage of the debate to present them as if I had just discovered something.

To strike a useful balance between familiar notions of markets and what the state trusts specifically might teach us, let me bypass the bulk of what I take to be the normal range of discussion in this area and focus instead on two elements peculiar to the trust lands that seem to have special relevance to adapting general market theory to land management in what continues to be a distinctly political arena and an imperfect marketplace to boot: the emphasis in trust law on thinking in terms of a trust portfolio—particularly the importance in trust lands management of thinking about permanent funds—and the trust lands experience in program funding as a possible antidote to current thinking about sources of funding for Forest Service programs.

The Portfolio. The most interesting trust element, in my opinion, is the idea of the trust as a portfolio of assets. I believe that the Forest Service is deeply hard-wired against thinking in portfolio terms. Putting the portfolio mentality and some standard trust activities into the Forest Service mental hopper would considerably enhance the agency's capacity for adaptive responses to local conditions and priorities.

The Forest Service seems to tie its own hands by thinking fairly consistently in terms of managing acres—often sacred acres—that are entrusted to it. For work that I am doing on conservation land acquisition, I recently reviewed in some considerable detail several of the forest-to-park land-transfer battles of this century, starting with the shifts that occurred with the passage of the National Park Service (NPS) Organic Act in 1916, continuing through the Roosevelt reorganization of monuments and battlefields, and capped probably by the protracted dispute over the evolution of Grand Teton National Park. (The Grand Teton story is probably the most familiar, but it is repeated throughout the country on many occasions [Connick and Fairfax draft].) I am deeply aware that the Forest Service has fought the NPS for "its" sacred acres for most of this century and that it regards repositioning its acres, ignoring some, or emphasizing others as some kind of defalcation.

This mind-set is not at all a part of the state trust lands experience. In Washington State, by comparison, the trust lands managers view the lands they hold as a part of their trust portfolio. Perhaps because Washington State is not one of the two trust organizations that manages the land and the permanent fund, I think it is fair to say that the state DNR considers the money to be a less important resource than the land. This kind of thinking causes some amusement when financial managers, auditors, and other private firms make presentations to trust lands managers, asking, "Why so much focus on thousands of acres that lose money when you could focus on the permanent fund and radically increase the funding for the schools?" The trust lands managers do not consistently "get it." But the state trust is happy to reposition itself off of areas that are politically difficult to manage and/or environmentally sensitive—it trades them or sells them to the state park program. The only constraint is that the trust be compensated.

The Forest Service does not easily consider repositioning itself. Nor does it think easily of investing in the development of resources that will produce returns—or of *not* investing in resources that will not produce returns. Finally, it does not appear to think in terms of achieving acceptable levels of risk by spreading investment or techniques across resources, or spreading management techniques or investments across resources differently. One reason is quite clear—until relatively recently, no one has seriously considered that the Forest Service ought to think in terms of producing returns or spreading risks. The Forest Service way was right for every acre. The Forest Service could gain considerable mental flexibility if it opened its mind to thinking in terms of a portfolio rather than sacred acres to be managed according to a catechism. The portfolio concept is the first step toward identifying valuable resources to manage, analyzing opportunities and resources for local economic development, and experimenting with tools and techniques.

For some shock therapy, let me share with you some important trust lands resource management decisions. From the mid-1970s to the early 1990s, a period during which land prices were increasing many times more rapidly than inflation, the Idaho State Land Board reduced its agricultural land holdings from about 55,000 acres to about 6,000 acres. Its rental income declined accordingly, from about $475,000 per year to about $100,000 per year in the early 1990s. Proceeds from the land sales were placed in the permanent school fund. Idaho sold approximately 90% of its agricultural lands, lost only 79% of its agricultural revenues, and produced three to six times its previous revenues (Souder and Fairfax 1996, 103–4).

Washington took a significantly different approach to its agricultural lands. Beginning in the late 1950s, the state decided to invest in irrigation to raise the value of selected lands. When the program started, Washington leased fewer than 500 acres of irrigated land. Today, the state leases nearly 34,000 acres of irrigated agricultural parcels. The investment in the conversion was approximately $10 million, and revenues increased from $0.50 per acre for dry land to $50–500 per acre for irrigated row crops, orchards, and vineyards.

Perhaps the most breathtaking trust lands programs of all involve the "transitional lands" programs in many states. The basic idea is that as towns grow out to meet trust holdings, the trustees increasingly engage in commercial real estate developments. Almost every state has managed at least some commercial properties—at a minimum, parcels that have come to the trust for tax losses or parcels that are astride a major roadway and are leasable for gas stations, warehouses, or similar facilities. Washington also has developed a "land bank" that temporarily holds land sales receipts pending purchase of replacement properties. The money eventually makes it into the permanent fund, but not before it is used to improve commercially viable properties. The grandpappy of the commercial developers is clearly the Arizona trust. It runs an Urban Lands Program that works to improve selected holdings for residential and—gasp!—golf course developments so that the state can enjoy a larger return than simply bare land value.

I mention these trust examples not to suggest that the Forest Service should go into real estate development but to suggest that the spectrum of what is done and doable on public lands out west is far broader than the Forest Service model suggests. Coming to grips with this broader spectrum of uses would enable the agency to think more in terms of a portfolio.

What would the agency get out of doing that? The portfolio perspective would help the Forest Service think critically about which resources in any region are the most important elements for management emphasis. The agency would be well advised to think both in terms of its own portfolio of assets in a region and in terms of the region's portfolio of assets. If the Forest Service is going to play an effective role in regional economic development, it must be able to see which of its resources are most worth managing and, not always the same, how national forest resources fit into a regional economic picture. Increasingly, the agency must have an eye on both those sparrows.

One example of the results of this perspective is so obvious that we overlooked it for years. It is instructive to point out that of twenty-two western states that own and manage trust lands, only four—Washington, Oregon, Idaho, and Montana—run serious timber programs. Yet, the Forest Service tries to manage timber in almost every state where its acres are located. Even allowing for differences in land location and quality, this disparity is striking. Because they are required to make money or break even, state trustees are considerably less enthusiastic about marketing timber in places where the Forest Service persists. This situation strongly suggests that the Forest Service could learn a lot about setting priorities if it experimented with thinking of its role and its resources in terms of agency and regional portfolios.

Even if the Forest Service is not ultimately responsible for producing returns, simply thinking about what is in its portfolio in terms of productive and unproductive assets—however defined—would be an important new view of agency lands. Thinking of different assets in a portfolio rather than acres to be managed ought to be of major value to the agency in setting priorities in a resource-constrained environment. It does not necessarily mean that the Forest Service would engage in a serious disposition program; however, it might help them develop areas of management emphasis. I am quite taken with the Washington model on irrigated lands. In my mind, it is equal to the program that BLM manages under the Recreation and Public Purposes Act—and the Forest Service could productively meditate on those examples.

Funding Mechanisms. The second element I want to discuss under the heading of a general shift from a scientific metaphor to a market one is the issue of funding trust lands programs. The state trust lands model is frequently the focus of undeserved attention, as many casual observers attempt to present the state trust lands as funded from revenues, hence embodying the most obvious feature of businesslike operations. As I have indicated, the truth is slightly different and potentially more useful.

What you *can* find in the trust lands context are numerous models for approaching that businesslike mode of operation without cutting the legislature completely out of the appropriations process. This Congress is understandably reluctant to do so. Therefore, the various models adopted in the several states for solving the same problem might be instructive to the agency—for reshaping individual programs or parts of programs. Or, Congress might experiment with some of the forms that states have used. Washington State's is perhaps the most interesting approach: Up to 25% of the revenues from both renewable and nonrenewable resources, including land sales, are deposited in the DNR's account for its operations on state trust lands. However, the funds must be appropriated before they can be spent. Moreover, any unspent funds are retained in an account by the trustees for subsequent expenditure. After several years, unexpended appropriations are distributed directly to the beneficiaries.

This pattern of operations funding mechanisms is not always as Randal O'Toole would prescribe. Moreover, Jon Souder and I were not able to discern any reliable correlation between the trustee's funding mechanism and policy outcomes. Nevertheless, funding mechanisms and processes are two areas in which reformers have recently focused a great deal of attention. State trust lands management agencies have adopted several variations on two dominant themes. This diverse experience ought to be part of the conversation when thinking about funding mechanisms for the Forest Service, national forests, and ranger districts. The states provide many models to examine.

Flexibility in Response to Institutional Diversification

If, as I have suggested, the future brings a period of diversification in conservation institutions that requires the Forest Service to tailor programs and partnerships to different regions and communities, then it probably is also true that the state land trusts have the most to teach the agency in terms of institutional flexibility. I emphasize two kinds of flexibility. First, as partnering with public and private organizations emerges as a commonplace method for coping with budget shortfalls, political pressures, and the demand for landscape-level decisions, the trust provides a wonderful vehicle for rapidly organizing entities to share access to, control over, and benefits from resources. A trust is basically a means for organizing title, control, and benefit. Understanding this flexible, adaptable model could be useful to the agency in many settings. Second, as the agency tries to work in ways that are more responsive to the peculiarities of location, it may allow itself to consider the possibility that management techniques perfectly adapted to one region or setting are inappropriate for another. One way to diversify the agency is to diversify its management tool kit. The state trust lands do similar things quite differently from the Forest Service, and on any specific topic, reference to their experience is probably a source of useful insights.

Regarding institutional diversification, our work suggests that many of the details of an agency's administrative setup do not matter much. When I first began working on trust lands I was quite anxious, as any good political scientist ought to be, to delve into differences in outcomes that might be attributed to the kind of commissioner (elected, appointed, or civil service); the kind of board (active, appeals only, or moribund); the funding mechanisms; and so forth. As we enter a period in which institutions are forming and changing rather rapidly, it may be comforting to know that none of that seems to affect policy outcomes. A lot of the theory—including, as I have suggested, how programs get their money—has limited discernible impact on outcome.

My recent work on conservation trusts has shown more generally that trusts are easy to establish. It is quite simple to set up a trust organization to manage lands and funds; organize a shared distribution of mitigation funds; and protect a habitat, endangered species, or a host of other purposes. Even in the most contentious situations, such as the unraveling of the Grayrocks Dam litigation or the Garrison Diversion, interested parties were able, within a matter of weeks, to establish fairly successful organizations to address mutual concerns (discussed in Guenzler and Fairfax forthcoming).

In *Land Conservation Through Public/Private Partnerships*, Eve Endicott (1993) wonderfully profiled a land acquisition transaction that involved the Forest Service, the Nature Conservancy, a conservation buyer, a seller, and fifteen people signing twenty-one documents. Another tale involves a Gallatin National Forest tract in which three separate foundations and the Montana Land Reliance raised the purchase price and held back mortgages on separate parcels of land that the agency intended to buy. These complex acquisition and management transactions beg for the clarity of the trust instrument.

The Forest Service participated in one of the most interesting cases we are studying: the *Exxon Valdez* Oil Spill Trustee Council, which with relatively little difficulty, melded six state and federal agencies into a group that has been astoundingly successful at spending almost $1 billion—not so simple a task as you might imagine. One of the handy things about a trust is that whereas the framers can control whatever they set out to control, the general principles and long familiarity of the trust provides a default position for things that the framers forget or fail to address. Because the Forest Service is also increasingly likely to get drawn into institutional settings that are partnerships (for example, to receive, manage, and expend damages and mitigation funds or to participate in community planning and consensus groups) it could be important to have this ready format up the agency sleeve.

As institutions diversify, it is helpful to know that the state trust lands managers have developed a tool kit that is sufficiently closely related to the Forest Service's own that there may be some fruitful options and overlap. The state trust lands managers do many of the same things that the Forest Service does: lease grazing lands, sell timber harvest rights, and sell rights to access and develop minerals. Because of the different mandate, they simply do it differently. What I am suggesting is a wrench hunt, that is, looking for tools that work in a particular set of circumstances—longer handle and a shorter, wider mouth. Some state trust land manager has probably tried it. It probably is more efficient and responsive to market imperatives than "the Forest Service way" and probably is worth considering.

The Forest Service ought to be looking for options for dealing with expiration, renewal, transfer, and improvements on grazing leases; consequences for different approaches to subleasing, resource and land appraisal, and fee-setting schemes; and different ways to structure payment schedules on oil and gas leasing. Many of these options are detailed in *State Trust Lands* (Souder and Fairfax 1996) or in numerous compilations that the state trust lands organization puts together. Sometimes, the agency might consider adopting a different approach. Other times, it might be useful to know that the lease terms that lessees insist would put them flat out of business if adopted by the federal government run without a hitch on state lands. State lands managers, for example, charge fair market value for telecommunications sites, actually run a grazing *leasing* program, and have devised many ways to charge for recreation.

A few examples of user fees will have to suffice. First, although the management of trust lands public minerals on state lands is deeply colored by federal categories and concepts, they all charge for all mineral resources extracted: hard rock and energy minerals. Second, several states have extensive programs for not only managing water quality but also charging for water that arises on state lands. Montana and Colorado, for example, have made extensive efforts to gain control over water put to use on state trust lands. Normally, those rights are filed for by the lessee, which then limits the marketability of the lease. The Board of Land Commissioners (BLC) in Colorado is applying procedures in oil and gas leasing to groundwater management on a few parcels where they cannot get title outright. The state charges a 12.5% royalty on water sales for nonutility use and 10% on water used by public utilities. The Colorado program is a small one—approximately 30,000 acre feet are involved. However, the experience is worthy of consideration.

Probably of more importance, given the direction commodity development on the national forests appears headed, the state trust lands managers have taken various approaches to recreation access. Their experience could broaden the Forest Service's thinking on the subject. The Forest Service lands are, as I understand it, generally open with unrestricted access to recreationists. Most state trust lands managers surrounded by federal lands under those circumstances simply adopt the federal posture; because the state parcels are not separately fenced, any other stance would not be prudent. The trust lands run into controversy where they are surrounded by private lands that are posted. In that context, the state lands are the "public" lands, and the trustees are under pressure to provide access for general public recreation and hunting. And, of course, they encounter the opposite pressure from the lessees, who want the state lands closed.

The interesting issue is how states have attempted to gain some recreation returns in areas where they are under pressure to provide general access and it is not efficient to collect fees at a gate. Four different approaches are taken. The first is what we might call the Forest Service approach: Except where a lease specifically allows the lessee to exclude recreationists, the state simply allows free recreation access. This is most typical. However, three other approaches are worth considering. At the other extreme, some states simply close their lands to recreation. This option would allow the state to make a recreation lease with either an existing lessee or a supplemental lessee, who could then manage the recreation use on the site and pay an agreed-upon return to the trust. This policy occasionally is implemented when the trustee learns that an agricultural or grazing lessee is coincidentally leasing recreation access to the state parcel. Typically, the state will rewrite the existing lease to include a charge for recreation access or release the parcel for recreation uses to another lessee.

Montana undertook a study to learn how to maximize returns for recreation access to state trust lands. It had been charging $5 for an unlimited number of annual recreation permits issued for state lands. Following the recommendations of an economic consultant, the state adopted a policy of charging $35 for a restricted number of annual recreation permits. This allows the state, if sales goals are met, to maximize returns while allowing only 12,000 recreationists access to trust lands. Colorado took a different approach to hunting access by cooperating with the State Division of Wildlife. The trustees allow the division to specify up to 500,000 acres of trust lands that will be opened annually to hunting. For this access, the Division of Wildlife pays the trust $500,000 per year. No camping is allowed on any of the state trust lands, and they are open only during hunting season. Colorado receives approximately the same annual income from hunting access as Montana but allows much broader public access.

As the Forest Service enters an era in which recreation access is increasingly controversial and potentially increasingly profitable, the states' experiences with access and returns policy is important. At a minimum, they ought to suggest that some things now considered in many circles to be politically impossible or otherwise unthinkable are neither. The states have broad experience in several recreation programs that ought to help the Forest Service generate alternatives.

Trustees have developed a variety of public involvement programs that would enrich the exercise for the agency and perhaps build better connections with increasingly important local and regional publics. I find most interesting the developments undertaken in that area by the EVOS program. It evolved from and presently is way beyond what you might characterize as the "notice and comment" involvement that typifies the federal agencies. Whereas most state trust management organizations have stayed fairly close to the Forest Service example on public involvement—that is, protective of their authority to make decisions and therefore clear that they are seeking advice and "input" from the public, not actual participation in management—several of the land conservation trusts I have been studying are not so constrained.

The EVOS trust (discussed in detail in Guenzler and Fairfax forthcoming) is particularly relevant because the programs are responsive to the growing emphasis on community participation and because the Forest Service plays a major role in the EVOS council. The EVOS staff has been aggressive about devising ways to involve affected groups and individuals as deeply as possible in EVOS programs. Although not all of the EVOS efforts are effective or efficient, they provide a small library of tools and concepts available for experimentation.

This elaborate program for involving villagers in the spill-affected areas is particularly relevant. It has evolved into a self-conscious effort to develop routines for bringing local ecological experience and agency science together. The program has self-consciously adapted programs initiated by Alaska Native Science Commission (ANSC). The founding of the new group was supported with funding from the National Science Foundation. Initial goals of the ANSC were to focus research on issues that were of interest to native communities and involve them in research design; to integrate traditional knowledge into research and science, facilitating native participation and training for native young people; and to ensure that results of research were usefully disseminated to native villages while protecting their cultures and their intellectual property. Some of these expectations are reasonable in the context of cross-cultural and native subsistence issues. The program also provides an excellent start toward a more local or regional, interactive approach to forest management. In the now-familiar areas of soliciting comments and participation in planning processes, the EVOS program goes beyond what has become routine for the agencies and is worth study for that reason alone. But in the field of local involvement in science and management, EVOS is truly a pioneer, developing both programs and protocols for working with local publics in the management of resources. The Forest Service ought to be learning from programs in which it is already a key participant.

In all these areas, the Forest Service has much to learn from the state trust lands managers. The agency likes to think of itself as the touchstone for an efficient, effective public agency and the best resource management institution in the United States and the world. From the perspective of a trust lands student, I see a slightly different story: The best predictor of bad trust lands management, particularly in the grazing context, is the presence of extensive federal lands nearby. When the federal agencies establish the expectations, it is hard for the states to transcend the culture and the rate structure designed to benefit the established commodity users. This reality is worth consideration when contemplating what Smokey could learn from the states.

THE FUTURE: TRUSTS AS TOOLS

Americans are rethinking ideas about federal government and its relationship to science, markets, states, local government, and private groups. Because the Forest Service embodies the values and assumptions of the model now faltering, it is particularly important for them to think comprehensively about new tools, vocabularies, and stories about what they are doing and why. Trust principles and the experience of the state trust land managers have much to teach in this context because the broad social changes under way emphasize the core principles and virtues of the trust: a market rather than a scientific metaphor of organization, an enormous flexibility and adaptability in the face of institutional diversification, and a relationship to the courts that is significantly different from the one defined in the APA.

Therefore, as we look for ideas and experience to revitalize the long moribund vocabulary of public resource management, it is appropriate that trust principles should occupy a significant place in the discourse. Thinking in terms of portfolios, undivided loyalty to specified beneficiaries, and expertise as a goad to higher standards of performance rather than deference could radically reorder the Forest Service's pantheon. For example, because of its emphasis on accountability to clearly defined beneficiaries, the trust would provide an especially sharp razor in addressing the question of what uses and users should be subsidized and which should pay their own way or produce a return to the treasury.

However, looking to the trust lands for experience and evaluation of different tools for such activities as basic accounting, allocating and reallocating grazing leases, appraising improvements, and running a public involvement program is probably more attractive to the agency and likely to be more productive in the near term. We do not have to speculate or theorize wildly to explore the possible impacts of proposed reform approaches; much of what is considered progressive and/or impossible to achieve in the Forest Service arena is standard operating procedure somewhere on state trust lands. Some of the institutional and mechanical variation does not seem to make much difference in terms of policy. Those data are useful. Moreover, simply understanding what the traffic would allow would strengthen the Forest Service's hand in revising programs. Many procedures that the agency's constituents complain would put them out of business are routinely accepted by the same operators dealing on state lands. The trust lands are not perfectly managed, and trust principles are not a silver bullet. But they do give a library of experience to those who are looking for alternatives to the present Forest Service approach.

Trust principles will not transform the agency into a perfect organization for the new millennium. However, the Forest Service is presently hoisted on several petards to which trust principles are particularly responsive. The state trust lands are an important source of insight for reformers of all persuasions.

ENDNOTES

[1]Causes for this fall from grace are numerous and complex. Characteristically, Randal O'Toole emphasizes that the agency is no longer making money: In the 1990s, the agency lost approximately $2 billion annually by managing the same number of acres that it managed at a profit when *Newsweek* became nearly breathless with admiration in the early 1950s. O'Toole also notes the centralization of the agency that began in the 1970s as an additional contributing factor. Others have focused attention on the standardless multiple-use mandate that grants the agency discretion almost without fetter, but leaves it unable to define its priorities, to do or defend any particular action as comporting with a mandate. (The best discussion of the multiple use act is still that by McCloskey [1961].) Conversation nevertheless seems to focus on demoralization in the ranks—the agency lacks a "vision" of its mission, and its employees are individually and collectively buffeted from one witless paper-pushing exercise to another (Forest Options Group 1998; see also Fairfax 1980).

[2]It is common to observe that reluctance to change policy stems in part from the fact that the protagonists are so well paid off that none of them has much

incentive to consider alterations (Leshy 1984). Sax (1984) makes the same point in the same volume, observing that the public lands constitute a "mature" policy system in which all the relevant interests are paid off.

[3]Although as Schiff (1960) demonstrates, the agency is a fickle friend of scientific method. Newly designated former chief Jack Ward Thomas made the radical suggestion that the troops could improve their stature simply by telling the truth.

[4]I refer, of course, to *Mountains Without Handrails* (Sax 1980) and the whole array of Sax's parks works, discussed in Fairfax (1998).

[5]This is not to say that the emphasis on local decisionmaking disappeared either from the politics or the rhetoric of American life. However, after the defeat of William Jennings Bryan in 1896, the effect of the small town mystique receded as the federal government emerged as the dominant force in public lands and most other policy arenas (discussed in Raymond and Fairfax 1999).

[6]See *Camfield v. United States* (167 U.S. 518 [1897]), which is not an interpretation of either property clause but in fact a nuisance case: As proprietor, the federal government has a right to protect itself against nuisance.

[7]This fundamental notion of separation of powers is manifest in the Constitution but is even more deeply rooted. In *Second Treatise of Civil Government*, Locke (1948) observed that legislatures "neither must nor can transfer the power of making laws to anybody else, or place it anywhere but where the people have."

[8]Compare *United States v. Grimaud* (220 U.S. 506 [1911]) with *Lochner v. New York* (198 U.S. 45 [1905]). Lochner named the era of judicial hostility to federal programs and is the subject of an enormous literature. A good place to start is Sunstein (1987).

[9]The President would add a sufficient number of judges to the Court to ensure a protective majority for his programs. From whence cometh the famous "switch in time that saved nine."

[10]A key element of that role was territorial expansion—particularly under the authority of the Weeks Act—to provide the CCC "boys" with places to work. However, the Forest Service lost on two of its most ambitiously aggrandizing projects: the quest to manage the unreserved, unentered public domain as grazing districts, and its Copeland Report—a proposal for Forest Service acquisition of 224 million acres of private timber land (discussed in Connick and Fairfax draft).

[11]The passage of the statute was not, however, merely a clarification of emerging standards, although it was clearly that; nor was it simply an effort to tidy up unevenness in administrative practice. It was both an acknowledgment that depression-era agencies would not go away, and a victory for business interests threatened by the New Deal and thwarted in their efforts to roll back the growing role of federal regulators in their affairs. If the business community could not undo the "Roosevelt Revolution," at least they could "reduce the power of regulatory agencies by increasing their own procedural rights" (Horwit 1994, 141).

[12]Key cases include *Lujan v. National Wildlife Federation* (110 Sup. Ct. 3177 [1990]) and *Lujan v. Defenders of Wildlife* (112 Sup. Ct. 2130 [1992]).

[13]The key and most discussed case in this area is *Chevron v. NRDC* (467 U.S. 837 [1984]).

[14]But note the current efforts by Mary Landrieu and others to mildly reactivate the Land and Water Conservation Fund in return for allotting 27% of OCS (*Greenwire* 1998c).

[15]In part, it is important to acknowledge, this institutional diversification is the product of longstanding reform efforts, and the goal of many federal programs, in conservation and other arenas. On the eve of the civil rights movement, it is important to recall, state government did not seem to have much to offer beyond a tawdry cover for racial segregation. Three decades and a growing number of intergovernmental transfers of financial resources and authority later, the states and some localities are sufficiently resurrected to challenge federal authority in a growing number of arenas. The Air Quality Act, the Water Quality Act, and environmental era programs too numerous to mention are among those that were designed in part to enhance the capabilities of state and local government. The standard literature is discussed at length in Fairfax (1982).

[16]For example, oil and gas conservation and pool unitization requirements were first defined under state law. Those state laws have never been displaced by federal enactments, and almost all such programs are state defined, state run, and operative on federal lands (Fairfax and Yale 1987, 74). For a brief description of most relevant mineral leasing programs, Section 2 of Fairfax and Yale is still a good place to start. The federal land management agencies have for the most part relied upon state standards and capacities for enforcement of air and water pollution on federal land (see also Donahue 1996).

[17]The division in the environmentalist side of the house was probably at one time a matter of socioeconomic class: Joe Sax described "the distinguished New York lawyer and fly-fisherman [lying] by the side of a stream contemplating the bubbles, while the factory worker roars across the California desert on a motorcycle" (Sax 1980, 48).

[18]Trust law is of sufficient complexity and importance that it is one of the legal fields in which members of the bar periodically compile recent case law and commentary and "restate" the basic principles in an allegedly concise, readable format. Trust law generally is on its second restatement—published (adopted and promulgated) in 1959 under the umbrella of the American Law Institute (ALI). Significantly, the definition of what constitutes prudence has changed in such important ways that the ALI has recently published a third restatement of trust law covering only the issue of prudence (see Restatement [Third] of Trusts SS 277 [Prudent Investor Rule]).

[19]It also requires them to do things quite different from the Second Restatement, and long prior. Under the earlier construction, trustees were given a list of investments that were approved. Trustees investing in those listed investments were "unquestionably" acting prudently (Souder and Fairfax 1997, 180).

[20]Jon Souder and I have planned for many years to pursue this precise issue. One interesting way to get a handle on the issue is to observe problems auditors have encountered in federal government expense-reporting procedures where

the agencies are involved in the *Exxon Valdez* Oil Spill (EVOS) Trust. Because EVOS is a trust and the Forest Service is among its subcontractors and trustees, the Forest Service books are subject to a standard audit. The auditor is not impressed and has trouble following expenses, allocation of time to projects, and a host of minor details. The Forest Service could use EVOS experience for lessons on reputable bookkeeping, if such there be (discussed in Guenzler and Fairfax forthcoming).

[21]Regarding standing, Idaho is a pesky jurisdiction that does things differently and fouls up any generalization. Idaho courts define beneficiary narrowly and take a narrow approach to citizens who can sue to vindicate trust principles (see *Selkirk-Priest Basin Association, Inc., v. State of Idaho,* 96.11 ISCR 431 [1995]). Similar efforts in the grazing area have also been turned back (see a long line of cases named *Idaho Watershed Project v. State Board of Land Commissioners,* starting with CV 94-1171 [1994]).

[22]The state pointed to the growing markets for conifer bough sales, pole sales, mushroom-harvesting leases, and sales of small-diameter timber that did not appear promising until recently (see State Respondent's Brief in Opposition, at 29, citing FRP, Appendix A at 18 and 19 in *Okanogan County v. Belcher,* No. 95-2-00867-9, Sup. Ct. for Chelan County; discussed in Souder and others 1998).

REFERENCES

Bogert, George Taylor. 1987. *Trusts,* 6th ed. St. Paul, MN: West Publishing Co., 348–9.

Carson, Rachel. 1962. *Silent Spring.* Boston, MA: Houghton Mifflin.

Connick, Sarah, and Sally Fairfax. Draft. *Federal Land Acquisition for Conservation: A Policy History.*

Cowart, Richard, and Sally Fairfax. 1988. Public Lands Federalism: Judicial Theory and Administrative Reality. *Ecology Law Quarterly* 15: 375–476.

De Voto, Bernard. 1955. *The Easy Chair.* Boston, MA: Houghton Mifflin.

Donahue, Debra L. 1996. The Untapped Power of Clean Water Act Section 401. *Ecology Law Quarterly* 23: 201–301.

Endicott, Eve (ed.). 1993. *Land Conservation through Public/Private Partnerships.* Washington, DC: Island Press, 199–202.

Fairfax, Sally. 1978. A Disaster in the Environmental Movement. *Science* 199: 743–48.

———. 1980. RPA and the Forest Service. In *A Citizen's Guide to the Resources Planning Act and Forest Service Planning,* edited by William Shands. Washington, DC: The Conservation Foundation.

———. 1982. Old Recipes for New Federalism. *Environmental Law* 12: 945–80.

———. 1987. Interstate Bargaining over Revenue Sharing and Payments in Lieu of Taxes: Federalism As If States Mattered. In *Federal Lands Policy,* edited by P. O. Foss. Westport, CT: Greenwood.

———. 1997. Grazing Leasing on State Trust Lands: Four Current Cases. In *Environmental Federalism*, edited by P. J. Hill and Terry Anderson. Lanham, MD: Rowman and Littlefield.

———. 1998. The Essential Legacy of a Sustaining Civilization: Professor Joseph Sax on the National Parks. *Ecology Law Quarterly* 25(3): 385.

Fairfax, Sally, and Carolyn Yale. 1987. *Federal Lands: A Guide to Planning, Management, and State Revenues.* Washington, DC: Island Press (and references cited therein).

Fairfax, Sally, Louise P. Fortmann, Ann Hawkins, Lynn Huntsinger, Nancy Peluso, and Stephen Wolf. 1999. The Federal Forests Are Not What They Seem: Formal and Informal Claims to Federal Lands. *Ecology Law Quarterly* 25: 630–46.

Forest Options Group. 1998. *The Second Century Report: Options for the Forest Service 2nd Century.* Oak Grove, OR: Thoreau Institute. http://www.ti.org/2c.html

Forest Service Employees for Environmental Ethics. 1996. *Who Says Money Doesn't Grow on Trees?* Occasional paper. afseee@afseee.org (accessed July 29, 1999).

Greenwire. 1998a. Poll: Americans Support Science but Don't Know It Well. 4 (July 2).

———. 1998b. Western Govs: Leaders Adopt Resources Manifesto. 5 (June 22).

———. 1998c. Royalties: Proposal Would Give States Offshore Revenue. 9 (July 22).

Guenzler, Darla, and Sally Fairfax. Forthcoming. *Conservation Trusts: Institutional Design for a New Era in Land and Resource Conservation.* Lawrence, KS: University Press of Kansas.

Hays, Samuel P. 1959. *Conservation and the Gospel of Efficiency.* Cambridge, MA: Harvard University Press.

Hess, Karl. 1993. *Rocky Times in Rocky Mountain National Park.* Niwot, CO: University Press of Colorado.

High Country News 1998. The Old West Is Going Under (Special Issue). 30(1).

Horwit, Steven. 1994. Judicial Review of Regulatory Decisions: The Changing Criteria. *Political Science Quarterly* 109(133): 141.

Kuhn, Thomas. 1962. *The Structure of Scientific Revolution.* Chicago, IL: University of Chicago Press.

Laitos, Jan. 1998. *The New Dominant-Use Reality on Multiple-Use Lands.* Keynote Address at the Rocky Mountain Mineral Law Institute, July 22, 1998.

Leshy, John. 1984. Sharing Federal Multiple-Use Lands: Historic Lessons and Speculations for the Future. In *Rethinking the Federal Lands*, edited by Sterling Brubaker. Washington, DC: Resources for the Future, 235–74.

Locke, John. 1948 [1704]. *Second Treatise of Civil Government.* Oxford, U.K.: Blackwell.

Lowi, Theodore. 1969. *The End of Liberalism: Ideology, Policy and the Crisis of Public Authority.* New York: Norton.

McCloskey, Michael. 1961. The Multiple Use Sustained Yield Act of 1960 (Note and Comment). *Oregon Law Review* 49(41): 49–77.

Meyer, Norman. 1994. The Dance of Nature: New Concepts in Ecology. *Chi-Kent L. Rev* 69:875.

Nelson, Robert H. 1995. *Public Lands, Private Rights: The Failure of Scientific Management.* Lanham, MD: Rowman and Littlefield.

O'Toole, Randal. 1994. Pork Barrel and the Environment. *Different Drummer* 1 (1).

Peffer, Louise. 1951. *The Closing of the Public Domain: Disposal and Reservation Policies, 1900–1950.* Stanford, CA: Stanford University Press.

Popper, Frank. 1988. A Nest-Egg Approach to the Public Lands. In *Managing Public Lands in the Public Interest*, edited by Benjamin C. Dysart, and Marion Clawson. New York: Praeger, 87.

Public Lands News. 1997a. Does Shea Hold Key to State Oil and Gas Role? 22(19): 9–10.

———. 1997b. Wyoming, IPAA Think Alike on Oil Royalty. 22(22): 6–7.

Raymond, Leigh, and Sally Fairfax. 1999. Fragmentation of Public Domain Law and Policy: An Alternative to the "Shift to Retention" Thesis. *Natural Resources Journal* 39: 649–753.

Rodgers, William. 1994. Adaptation of Environmental Law to the Ecologists' Discovery of Disequilibria. *Chi.-Kent Law Review* 69: 887.

Sax, Joseph. 1970a. *Defending the Environment.* New York: Knopf.

———. 1970b. The Public Trust Doctrine in Natural Resources: Effective Judicial Intervention. *Michigan Law Review* 68 (1):417–566.

———. 1980. *Mountains without Handrails.* Ann Arbor, MI: University of Michigan Press.

———. 1984. The Claim for Retention of the Public Lands. In *Rethinking the Federal Lands*, edited by Sterling Brubaker. Washington, DC: Resources for the Future, 125–28.

Schiff, Ashley. 1960. *Fire and Water: Heresy in the U.S. Forest Service.* Cambridge, MA: Harvard University Press.

Shapiro, Martin. 1983. Administrative Discretion: The Next Stage. *Yale Law Journal* 92(1487).

———. 1988. *Who Guards the Guardians.* Athens, GA: University of Georgia Press.

Souder, Jon, and Sally Fairfax. 1996. *State Trust Lands: History, Management, and Sustainable Use.* Lawrence, KS: University Press of Kansas.

———. 1997. Arbitrary Administrators, Capricious Bureaucrats and Prudent Trustees: Does It Matter in the Review of Timber Salvage Sales? *Public Land and Resources Law Review* 18: 165–212.

Souder, Jon, Sally Fairfax, Theresa Rice, and Lawrence MacDonnell. 1998. Is State Trust Land Timber Management "Better" than Federal Timber Management? A Best Case Analysis. *West/Northwest Journal of Environmental Law and Policy (Hastings)* 5(1&2): 1–43.

Souder, Jon, Sally Fairfax, and Lawrence Ruth. (1994) Sustainable Resources Management and State School Lands: The Quest for Guiding Principles. *Natural Resources Journal* 34: 271–304.

Stewart, Richard. 1975. The Reformation of American Administrative Law. *Harvard Law Review* 88(2): 1669–1813.

Sunstein, Cass. 1987. Lochner's Legacy. *Columbia Law Review* 87(2): 873–919.

U.S. Commission on Fiscal Accountability of the Nation's Energy Resources. 1982. *Report of the Commission* (January).

U.S. GAO (General Accounting Office). 1996. *Public Timber: Federal and State Programs Differ Significantly in Pacific Northwest.* Report to the Chairman, Committee on Resources, House of Representatives. GAO/RCED-96-108. Washington, DC: U.S. General Accounting Office.

Uram, Robert. 1998. *Trends and Developments in Cooperative Federalism and the Regulation of Coal Mining.* Presented at the 12th Annual Developments and Trends in Public Land, Forest Resources, and Mining Law Conference of the American Bar Association Section of Natural Resources, Energy, and Environmental Law. March 6–7, 1998, Scottsdale, AZ.

Wherhan, Keith. 1992. The Neoclassical Revival in Administrative Law. *Administrative Law Review* 44: 567–72.

Wiebe, Robert. 1967. *The Search for Order—1877–1920.* New York: Hill and Wang.

Worster, Donald. 1977. *Nature's Economy: A History of Ecological Ideas.* New York: Cambridge University Press.

Discussion

State Trust Lands Management

Randal O'Toole

In the 1952 *Newsweek* article cited by Sally Fairfax, the magazine noted that "Most congressmen would as soon abuse their own mothers as be unkind to the Forest Service" (*Newsweek* 1952). The magazine credited the agency's popularity to the fact that it was "the only major government branch showing a cash profit" and "one of Uncle Sam's soundest and most businesslike investments." In turn, the author of the article added, "The Forest Service owes much of its phenomenal efficiency to two policies: decentralization and cooperation with anyone who will cooperate."

Today, the agency is highly centralized, and polarization has replaced cooperation. Forest Service abuse is a favorite sport on Capitol Hill. Far from earning a profit, the agency loses $2 billion per year managing the national forests. How did this once-proud organization, widely regarded as a model of excellence in government, fall so far? One possible answer is that the Forest Service somehow lost the virtues that made it so popular in the 1950s. But a more subtle response is that the Forest Service's own highly centralized institutional structure, which seemed to work in the 1950s, limited its ability to adapt to changing public demands and tastes. For this reason, it is useful to examine alternative institutional designs.

Sally Fairfax and her colleague Jon Souder have done the conservation world a major service in pointing out an important alternative to the standard progressive-era structure that is best exemplified by the Forest Service. Souder and Fairfax's (1996) detailed analysis of the state trust lands demonstrates that these lands operate under an institutional structure that is very different from, and in some respects superior to, the Forest Service, the Park Service, and other federal agencies.

RANDAL O'TOOLE is director of the Thoreau Institute and a visiting lecturer at Yale University.

No one, least of all Fairfax and Souder, thinks that state trusts are perfect. Nor does the work of Fairfax and Souder suggest that transferring federal lands to the states, without the trust formula, will lead to any improvements in on-the-ground management. What their work shows is that, in thinking about the future of the Forest Service and other federal resource agencies, we do not have to confine ourselves to the kind of institution that Gifford Pinchot and his associates designed nearly 100 years ago.

In 1995, I did my own analysis of state land and resource agencies, an analysis that was not as detailed as Fairfax's review of state trusts but that covered about seven times as many agencies (O'Toole 1995). I found that the states have taken three very different approaches to their resources. State forest agencies, including but not limited to the state trust lands, are generally operated as for-profit organizations. The agencies' managers are expected to produce more revenue than expenses, and about two-thirds of them do. At the other extreme, state park agencies tend to be heavily subsidized. Although New Hampshire and Vermont are notable exceptions, on the average, user fees cover less than one-third of the costs of managing the parks. Taxpayer subsidies amount to about $70 per acre per year, nearly four times as much as in federal parks. And in the middle of the pack are the state wildlife agencies. Until recently, they have been run as not-for-profit organizations, operating almost exclusively on their own user fees. Although the states have received federal dollars in the form of the Pittman–Robertson and Dingell–Johnson funds, these have been, in effect, merely more user fees. These funds are derived from taxes on hunting and fishing equipment and are distributed to the states according to the number of hunting and fishing licenses sold. In recent years, concerns about nongame wildlife have led many states to dedicate or appropriate certain tax revenues to the wildlife agencies, but as of 1995, they still amounted to less than 20% of state wildlife agency budgets.

Each of these three agency paradigms—for-profit, not-for-profit, and subsidized—has its virtues and pitfalls. But as a general rule, state and federal resource managers will find ways to spend all of the appropriations they are given plus all of the receipts they are allowed to keep. This is not to suggest that agency managers are in any way venal or corrupt. It is only that the number of things that can be done to manage resources is almost limitless. Whereas managers can always find some benefit from spending more money, without some safeguards, there is no guarantee that the money they spend is really the most appropriate place for society to put its resources.

So, what are the appropriate safeguards? One possibility is the trust mechanism described in detail by Fairfax. Such a mechanism will work best when there is a consensus that profitability, or at least self-sufficiency, is an appropriate goal for public resources.

A second necessary condition is that the beneficiary must get a substantial share of its budget from the trust so that it will have incentives to monitor the agency managing it. The poor performance of several state trusts can be traced to the fact that schools get so much of their money from general tax revenues that they ignore the trust lands. This situation suggests that counties, which traditionally have received one-quarter of national forest receipts, would make better beneficiaries than, say, the Social Security system, for which public land revenues would be practically invisible.

Even with these conditions, trusts are not a sufficient instrument for ensuring sound resource management. An additional safeguard must be provided in the form of a budgeting system that rewards good management and penalizes bad management. There are four possible ways for agencies to fund resource management: (1) appropriations out of general funds plus a share of gross receipts (an agency funded this way will have an incentive to lose money and to cross-subsidize unprofitable activities with profitable ones); (2) appropriations alone (an agency funded this way will have no incentive to make money and will tend to minimize receipts by encouraging resource users to substitute services to the agency for resource payments); (3) a share of gross receipts alone (an agency funded this way will have an incentive to set costs equal to receipts by cross-subsidizing unprofitable activities with profitable ones); and (4) a share of net receipts alone (an agency funded this way will have an incentive to maximize profits and minimize cross-subsidization).

The alternative of funding out of appropriations plus a share of net receipts is conceivable but not necessary, because an agency earning net receipts needs no appropriations. How would funding out of net receipts work? At its simplest level, "net" or profit equals receipts minus costs. Funding an agency out of net receipts would mean that, each year, the agency would report (subject to audit) its income and its expenses. The excess of income over expenses would go into the agency's account for expenditure in the following year.

Say that in one year, an agency spent $7 million and earned $12 million in revenue. Its net would be $5 million, and it would have that amount to spend the following year (possibly on top of funds carried over from previous years). Because the $7 million that it spent came from funds earned during a previous year, that means that there is $7 million in receipts that can be distributed to beneficiaries, returned to the government treasury, or spent in some other way.

Because net plus cost equals gross, an agency funded out of 100% of its net will tend to keep about half of its gross. (And an agency funded out of half its net will keep about one-third of its gross; funded out of a third of its net, it will keep about one-quarter of its gross; and so forth.) But the incentives created by funding out of net are very different from funding out of half of gross, because the latter encourages cross-subsidization, whereas the former does not.

Although I would argue that funding an agency out of its net receipts is necessary to ensure sound management, it is clearly no more sufficient by itself than the trust form. The trust mandate to preserve the corpus of the trust is just as necessary. At least one more safeguard is also needed to protect resources that cannot generate user fees.

If forests are to be turned into trusts, creating a nonmarket stewardship fund that is one of the beneficiaries of the trust can protect nonmarketable resources. This fund would be managed by an outside entity whose mandate would be to use the funds to protect and enhance nonmarket values on the national forests—and possibly on adjacent private lands. The trustee, which might be a nonprofit organization or an existing government agency, could use its funds to pay national forest managers to use or avoid certain practices that it deems to be beneficial or harmful.

My grand design for Forest Service reform, then, includes a series of checks and balances aimed at making management flexible and adaptive to new scientific findings and changes in public goals, minimizing bureaucracy, improving on-the-ground management, and ending polarization. These checks and balances include

- the legal form of the trust, which obligates managers to produce revenues for beneficiaries as well as to preserve the corpus of the trust;
- a board of trustees that will oversee the trust managers;
- trust beneficiaries that have incentives to monitor the trust;
- at least one trust beneficiary that has a mandate to use its revenues to protect and enhance nonmarket values on the trust lands; and
- a funding mechanism that rewards managers for earning a profit, thus discouraging cross-subsidization and excessive overhead.

How can such revolutionary reforms take place in a political system dedicated to incremental change? This question was asked by a group of some two dozen interest group leaders, policy experts, and Forest Service officials who met and corresponded throughout 1997 and 1998. The Forest Options Group, as it calls itself, went further and asked, "If we were the Forest Service's founders in 1905, but knowing what we know today, how would we design the agency?" (Forest Options Group 1998).

The group, which includes such people as industry attorney Steve Quarles, environmental advocate Andy Stahl, rancher Brad Little, forest supervisor Jim Furnish, and Sally Fairfax, specifically decided not to debate the mission of the Forest Service. Endless debates over the mission are ultimately fruitless, because any mission must necessarily be vague. The group also decided not to discuss on-the-ground management issues such as forest health or biodiversity conservation. Instead, the group focused on the governing and budgeting structures that create incentives and direction for on-the-ground managers.

The Forest Options Group considered various structures. Two alternative governance systems are making the forest supervisor report to a board of directors and turning the national forests into a forest trust, similar to the state land trusts. Alternative budgeting processes include funding a national forest out of its gross or net income, rather than out of tax dollars.

The group agreed that the only way to determine which structures work best is to test these alternatives on various national forests. So, the group has proposed five pilot tests:

1. Entrepreneurial budgeting: Fund the national forest out of its net income, while reserving some of the forest's receipts for nonmarket stewardship activities.
2. Collaborative management: Make the forest supervisor responsible to a collaborative board of directors.
3. Collaborative planning: Give a collaborative board the authority to design forest plan alternatives and to determine the selected alternative after appropriate public comment.
4. Forest trust: Create a board of trustees appointed by the secretary of agriculture and the governor of the state in which the forest is located and make the board responsible for preserving the corpus of the trust and for producing revenue for beneficiaries. The forest would be funded out of half of its gross receipts. Remaining receipts would be split between two beneficiaries, the counties and a nonmarket stewardship fund.
5. Gross receipts/rate board: Fund the forest out of its gross receipts and create a rate board to set equitable user fees.

For test purposes, all of the pilot programs must follow existing laws and regulations, but they would be relieved from following the Forest Service Manual or memo direction. Their budgets would be an open bucket, and most of the pilot-test forests could charge a broad range of user fees.

The Forest Options Group proposes that forests nominate themselves to become pilots. In the case of the collaborative pilots, the nominations would come from the self-selected collaborative boards. The secretary would be expected to select collaborative groups that represent a full range of interests. In addition to the five pilot tests described here, forests would be invited to design their own programs that might include a different combination of budget and governance structures.

Pilot tests 1, 4, and 5 would receive seed money during the first two years and a safety net to ensure that their budgets do not fall below half of their historic level. All pilot programs would report to an "office of pilot projects" instead of their geographic regions. This office would monitor the pilots and provide an interface between the pilot programs and Washington, DC.

The pilot tests would continue for five years, or longer if needed to determine which structures work best. Eventually, these programs could provide enough information to allow a reform of the Forest Service as a whole. The Forest Options Group recognizes that some structures might work best on some forests, whereas other forests would need other structures.

None of the suggested pilot programs is perfect. In particular, for political reasons the Forest Options Group does not propose to waive the National Environmental Policy Act (NEPA) and the National Forest Management Act (NFMA) for the pilot forests, which leaves them subject to all the foibles of the planning process. But at least they will be free to interpret the law for themselves and not required to follow the several thousand pages of manual provisions and memo direction. Ultimately, the pilot programs may show that alternative public involvement processes are far superior to NEPA and that alternative management systems are far superior to NFMA.

Many Forest Service employees, especially those closest to the ground, should welcome these pilot programs. The Forest Options Group expects that all of the pilot-test forests will in their own way improve land stewardship and management efficiency while reducing polarization. Some will reach some of these goals better than others. Although pilot scenario 4 comes closest to my preferred design, other group members have their favorites, and we all agree that only testing will determine which system will work best and help lead the agency into its second century.

CONCLUSIONS

The Forest Service and the national forests are in trouble. As the agency is beset with political and financial problems, the land is afflicted with ecological ones. These troubles are closely related and we will not be able to solve one without solving the other. Changing the agency's name, the department to which it reports, or the people who are at the top will do nothing for the forests. Instead, we need to change its financial and governance structures.

Examining state resource agencies provides many lessons for would-be Forest Service reformers. Merely transferring forests to the states will do little good since the states haven't proven to be uniformly superior resource managers themselves. But they have tried enough different systems to provide some hints of how national forests can be reformed.

My ideal reforms would be to turn each national forest into an individual federal forest trust, fund it out of its net receipts, and give a share of its gross receipts to a nonmarket stewardship fund. Other people prefer other governing and budgeting structures.

The only way to tell for certain which system works best will be to test proposed reforms on individual national forests. The Second Century proposal provides an opportunity for incremental revolution. Its pilot tests are incremental enough to be acceptable to Congress and most interest groups. Those tests lead to a process that could revolutionize the agency as a whole. Only such reforms of the Forest Service's governing and budgeting structures will lead to real improvements on land stewardship and public approval of this historic but controversial agency.

REFERENCES

Forest Options Group. 1998. *The Second Century Report: Options for the Forest Service 2nd Century.* Oak Grove, OR: Thoreau Institute. http://www.ti.org/2c.html.

Newsweek. 1952. June 2.

O'Toole, Randal. 1995. State Lands and Resources. *Different Drummer* 2(3): 4–23, 34–35, 42–43, and 54–56.

Souder, Jon, and Sally Fairfax. 1996. *State Trust Lands: History, Management, and Sustainable Use.* Lawrence, KS: University Press of Kansas.

7

Predicting the Future by Understanding the Past

A Historian Considers the Forest Service

Paul W. Hirt

As the twenty-first century begins, we have ample reason to ponder the past and future of America's national forests and the federal agency that manages them: the Forest Service. Both are largely products of this century, both evolved together through two main stages of historical development divided at midcentury, and both face a third and controversial new phase in their history as we begin a new millennium. Reviewing that history in the context of the current controversy over national forest management helps us understand how we got where we are today and provides clues to what we might expect in the next decade.

MY APPROACH AND INVOLVEMENT

As both a historian and a conservationist, I approach my task from two complementary angles: In looking to the past, I both examine the history of national forest management and evaluate our successes and failures in achieving environmental stewardship goals. In looking toward the future, I consider where we are likely to be in ten years (based on past experience) and offer a vision of where the agency might aim itself in the next decade—with the support of Congress, the executive branch, and the American public—to better achieve its stewardship goals.

PAUL W. HIRT is a professor of history at Washington State University.

By "stewardship goals," I mean management that minimizes biological harm and protects ecosystem health and diversity while providing various public goods and services, both market and nonmarket, in a balanced and sustainable manner. In general, this definition always has been the fundamental purpose of the national forests and the basic mandate of the Forest Service: protection of the forests while providing multiple uses under the principle of sustained yield. It is a wise mandate but one that has not been faithfully implemented over the past half century. Today, many people argue that this fundamental mandate is outdated and dysfunctional. In contrast, I suggest that the dysfunction is not in the goals themselves but in our failure to achieve them. This institutional problem is exacerbated by a narrow, commodity-oriented interpretation of "proper" forestry that has reigned since at least midcentury. Environmental protection, stewardship, and sustainable use remain a laudable mission for the Forest Service—if only we could agree on the terms and effectively implement them.

Most historians who write about the Forest Service acknowledge two major phases in the history of the agency and the management of the national forests: the "custodial era" of low-intensity management, which lasted from the beginning of the century to World War II, and an era of greatly expanded public use of the national forests, intensified resource development, and increased organizational growth of the Forest Service, which lasted through the 1980s. (For more information about the two main historical eras of the Forest Service, see Steen 1977; Clary 1986, 119–25; West 1992; and Hirt 1994, chapter 3.) In the 1990s, a dramatic shift in forest management came in response to several important developments: conflict over the accelerated national forest timber harvests of the post-World War II era, changing public values since the 1970s, increased costs and public subsidies required to remove timber from ever-more-remote and marginal forest lands in the 1980s, and an old-growth forest preservation movement that arose in the 1980s in response to the increasing scarcity of old growth. As a result, we have seen a two-thirds decline in timber harvests, significant reductions in new road construction, the downsizing of the agency's workforce, and a conspicuous shift to a new "ecosystem management" paradigm since 1990. The last decade of the twentieth century has clearly marked the beginning of a third era in national forest management (Hirt 1994, 296–7; see also the series of articles by Paul Mohai, Timothy Farnham, Jennifer Thomas, and Elise Jones [Symposium on Change 1995] and the series of articles on present trends and future directions for the Forest Service [*The Pinchot Letter* 1998, 11–19].).

Accompanying this dramatic period of change in management direction have been pitched battles between forest user groups, vituperative criticism of the Forest Service from all sides, convulsive internal conflict within the agency (viewed in part as an identity crisis), unprecedented congressional and judicial interventions in forest management, and numerous efforts to chart a path through the cross-fire to a better future. Resources for the Future's "A Vision for the Forest Service" (the conference that spawned this book) is one helpful path-finding effort in this time of transition.

The circumstances surrounding my initial involvement in national forest management help explain my interest and approach to the subject. During the 1980s, when the Forest Service was producing its first round of comprehensive forest plans under the auspices of the National Forest Management Act (NFMA) of 1976, I was living in Tucson, Arizona, coordinating public land–conservation efforts for the Grand Canyon Chapter of the Sierra Club. The Coronado National Forest, a collection of rugged mountain ranges surrounded by desert valleys in southeastern Arizona, produced its first-draft forest plan in 1982. It was one of the first forests in the nation to publish a plan under the NFMA forest-planning regulations. I formed a team of volunteers with expertise in a wide variety of fields, and we carefully reviewed and commented on the plan, working closely with the Coronado forest-planning team. Because this undertaking was the Coronado's first attempt at comprehensive land management planning, the initial forest plan and draft environmental impact statement (EIS) contained so many inadequacies that the agency produced a second, substantially revised draft in 1985 and resubmitted it for public comment. After a couple more years of evaluation and revision, the agency completed and published the Coronado National Forest Plan and Final EIS in 1987.

From my perspective as an interested outsider, the forest-planning process was complicated and sometimes frustrating, but ultimately effective in two key ways: It provided a comprehensive summary of current conditions on the national forest and assessed the contemporary management situation. For the first time, one single document assessed all forest resources together and integrated a discussion of problems, opportunities, and management objectives for the whole forest. Most important, the forest managers had to provide this document, free of charge, to interested citizens and to other state and federal agencies, soliciting their feedback and revising their analyses and objectives on the basis of those responses. The forest plan clarified options, compared management alternatives, and revealed to the public the agency's plans for the future of its lands in southeast Arizona. Even though the plan still had significant flaws and much of it was not subsequently implemented (for reasons I explain later), it seems to me that the process was valuable as a tool for facilitating communication, public disclosure, oversight, peer review, and greater honesty. Like peer review in academic publications, this process of public disclosure, feedback, and reevaluation exposed public forest management across the nation to the light of day. Regardless of the outcomes, I believe this element is important to democratic decisionmaking and a necessary foundation for learning and improvement.

But the results were not all positive. First, for most of the decade a very large proportion of the agency's human resources went into planning rather than actual forest management. Every hour and dollar spent generating paperwork were an hour and a dollar not available for on-the-ground accomplishments. Second, many of the plans and the data on which they were based remained deeply flawed; some were hardly worth the paper on which they were printed. Third, forest plans could and did often respond more to political pressures than to a rational calculation of what was best for the resources. Fourth, Forest Service monitoring reports in subsequent years revealed that plans were partially or lopsidedly implemented, mainly because plans tended to promise much more than the agency could deliver. In the case of the Coronado National Forest, one-third to one-half of the proposed accomplishments for the first decade remained unmet. The Coronado's fifth-year monitoring report, for example, blamed insufficient financial resources and personnel as the culprits, rather than acknowledging the real problem: an unrealistic, overly ambitious plan. For instance, the forest plan proposed to fix environmental problems associated with a century of overgrazing on the Coronado by more intensive management: more fencing, more water developments, better animal control, greater cooperation from rancher permittees, reseeding, control of soil erosion, and so forth. This plan required a lot of money and personnel—much more than the agency had budgeted and received for range management in the 1980s when it wrote the plan, and much more than it would receive in the 1990s. After only five years, the Coronado Forest planners had essentially left the plan behind and resorted to establishing annual "work priorities" based on each year's budget expectations (Coronado National Forest 1993). This result of forest planning has been common all across the nation.

In hindsight, the forest-planning process may be judged unsuccessful, because forest plans have not in fact been faithfully implemented. Words on paper do not have the power to determine the future; they cannot fix annual appropriations from Congress or control the market or compel human behavior or determine biological responses to management. Forest plans alone cannot solve inherited management problems, either. Plans, moreover, do not resolve public conflict over management decisions or encourage greater trust and acceptance of Forest Service authority. If plans are expected to accomplish these objectives, then the planning process will be judged a failure.

So, why do we bother? For the same reason we gather together to discuss the future of the Forest Service: because assessing management alternatives and charting a future course is valuable, even if the future remains unpredictable. Achieving conservation goals requires that we assess the current condition of our natural resources, understand how they got that way, identify desired future conditions, and chart a path to achieve them. Even though writing fifty-year management plans is tilting at windmills, it is better than the alternative—haphazardly stumbling forward, allowing whims, markets, or managerial bias, squeaky wheels, politicians, or power-brokers to determine how resources are allocated without reference to larger goals such as sustainability, social equity, and ecosystem health.

I, for one, would not like to return to the laissez-faire days of the nineteenth century, before Americans adopted natural resource conservation as a national policy. I would fight to the end to retain our bounty of federal lands (national forests, national parks, fish and wildlife refuges, Bureau of Land Management lands, and so forth), as would most Americans. And even though the land management planning process is desperately in need of reform, I would not return to the 1940s, when about the only planning taking place was over timber sales. Our task today is to learn from the past, so we can improve our efforts to manage our public forest lands in the future, for the values we cherish.

LESSONS FROM PAST EXPERIENCE

National forest management is incredibly complex. Only in the humblest sense can we say that we exercise any control over natural resources, their use, and their future condition. Since the inception of comprehensive land management planning in the late 1950s, the Forest Service has diligently attempted to manage the national forests according to a rationally coordinated vision of development and forest-restoration activities, supported by legal authority and scientific expertise. In fact, management of the national forests has instead been an *unending series of negotiations and compromises* among (1) Congress, the administration, and the courts; (2) Forest Service district offices, supervisors' offices, regional offices, and the Washington office; and (3) competing interest groups and public demands. I briefly address each of these factions, because we must understand the institutional environment of national forest management if we want to realistically reform it.

The Forest Service is a creature of the federal government—a tool wielded by federal policymakers and administrators to achieve specific social, economic, and political objectives, such as contributing to national timber supply, protecting watersheds of navigable rivers, providing wildlife habitat and recreation opportunities, or preserving biological diversity. But the federal government never speaks with one voice. Congress provides general statutory mandates, some of which are vague or conflicting, and annually appropriates most of the funds the Forest Service uses to manage the forests. These appropriations are often unbalanced, unpredictable, or tied to short-term, partisan political objectives. Al Sample, Randal O'Toole, Timothy Farnham, and I, as well as others, have shown how congressional appropriations have been poorly coordinated with the management plans developed by the Forest Service, making it difficult or impossible for the agency to implement its management objectives in a balanced, integrated manner, for at least half a century (O'Toole 1988; Sample 1990; Hirt 1994; Farnham 1995). Thus, the administrative branch and the legislative branch are constantly in a tug of war, pulling the Forest Service this way and that.

Needless to say, Congress itself never speaks with one voice, and neither does the administration. Shifts in party dominance often have led to significant shifts in congressional funding, legislation, and oversight (Sample 1990, chapter 2; Hirt 1994, chapters 5, 10, and 11). That will never change. Moreover, within the executive branch, the Forest Service often must negotiate or cooperate with other federal agencies that have divergent mandates: with the Environmental Protection Agency over water-quality concerns, with the Fish and Wildlife Service over endangered species, with the Bureau of Land Management over subsurface mineral rights, with the Bureau of Indian Affairs over treaty rights and obligations, with the Army Corps of Engineers over dams and water diversions, and so forth. To make matters even more complicated, the judicial branch occasionally exercises profound influence over Forest Service behavior, as shown in the famous Monongahela case that outlawed clearcutting in the 1970s (until NFMA passed) (LeMaster 1984; Wilkinson and Anderson 1985) or Judge William Dwyer's northern spotted owl decisions in the Pacific Northwest in the 1990s. (For a sensitive, journalistic view of these decisions, see Dietrich 1992.)

This fragmentation of the federal estate is inherent to the Forest Service, too. As an organization, the Forest Service has an administrative structure that tries to balance centralized and decentralized authority. The Washington office of the agency shares decisionmaking powers with its ten regional offices, which in turn share decisionmaking powers with 155 forest supervisors' offices, which in turn share authority with hundreds of district rangers' offices. (For an interesting and theoretically useful, if somewhat dated, study of fragmentation and integration within the Forest Service, see Kaufman 1960, 1967.) The forest-planning processes mandated in the 1974 Resources Planning Act (RPA) and the NFMA tried to integrate all these disparate levels of authority and knowledge with only minor success. National objectives reflected in the RPA plans rarely see full implementation at the local level, and the locally developed national forest plans generated under NFMA guidelines rarely get full institutional support from high levels of the agency. Coordinating the national RPA programs and the NFMA-based forest plans has been extremely difficult and not very effective (Sample 1990, chapters 1 and 10).

The Forest Service is fragmented horizontally as well as vertically. Besides the hierarchical layers of authority, strong differences exist among employees at each organizational level. Fisheries biologists, silviculturists, recreation managers, road engineers, and public relations experts have different interests, stakes, and professional orientations (for example, Farnham and others 1995a, 1995b). Although coordinated land management planning is enhanced by having diverse expertise among the planners, that same diversity is a double-edged sword; it also can stymie unity of purpose and efficiency. The Forest Service was a much more cohesive organization, efficiently pursuing clear management objectives, back in the days when decisionmakers were almost all foresters who had been trained at a handful of similar-minded forestry schools. NFMA's requirement for interdisciplinary planning, along with the proliferation of schools of natural resource science and environmental studies, has greatly diversified and fragmented federal land management organizations (Hirt forthcoming).

The challenges of coordinating this large semi-decentralized, diverse agency is complicated even further by the fact that it functions within a pluralistic, democratic society with a well-developed tradition of free speech, dissent, and organized interest group lobbying. Contrary to the declamations of conspiracy theorists on the right and the left, the historical record unequivocally shows that the U.S. government in general, and agencies like the Forest Service in particular, are indeed responsive to public pressures and changes in public values. As social scientists Jeanne Clarke and Daniel McCool noted, "The history of the Forest Service ... reveals a remarkable ability to sense changing public priorities and to adapt its mission to meet those demands" (Clarke and McCool 1985). (The issue of Forest Service responsiveness and resistance is discussed in Culhane 1981.) This characteristic responsiveness is both the blessing and the curse of democratic forms of governance, because although it guards against unresponsive insular authority, it also makes it difficult for agencies to follow through on long-term management strategies. Although those who govern the country may not always respond quickly or appropriately or in ways that many people would prefer, and although the ability of the "public" to pressure the government is not distributed uniformly among the population, natural resource decisionmaking is nevertheless influenced by a wide variety of public pressures, including interest group lobbies, employment conditions, public perceptions, media coverage, and swings in voter orientation. Once again, this situation makes for a complicated environment of negotiations and compromises that profoundly influences national forest management.

Of course, we social scientists must also remember that human perceptions and human institutions are not the only fickle forces of change that influence the Forest Service and other natural resource–managing agencies. Nature is as diverse, contradictory, unpredictable, and volatile as its human offspring. In the late 1980s, after a decade of raging conflict over the fate of roadless old-growth forest in the Siskiyou National Forest of Oregon, a ten-week-long 100,000-acre hot fire reduced a large chunk of the forest to charcoal (Egan 1990). Similarly, in 1980, after some fifty years of trail and campground development around Spirit Lake in the Gifford Pinchot National Forest in Washington, Mount St. Helens blew billions of tons of trees, ash, and sediment into the lake and covered tens of thousands of surrounding acres of beautiful forest land, turning paradise into a moonscape. Floods, windstorms, wildlife population fluctuations, fungi, insects, climate change, fire, and volcanic eruptions compound the complexity of our already complicated social institutions, making our efforts to rationally coordinate forest management dauntingly difficult, if not entirely futile.

None of these complications will go away. It would be naive to suggest that some coherent mission statement or authority could effectively guide the Forest Service down some predictable path in the future. However, the lack of a clearly marked trail does not mean we should avoid formulating a vision, setting policy guidelines, or striving for more coherence and consistency in national forest management. It simply means that we have to recognize our inability to fully determine outcomes and control the social or the physical environment. *Under such conditions, the most appropriate orientation for managers to take is humility, adaptability, and a conservative approach to resource management.* This theme runs throughout much of the literature on "new forestry" and "ecosystem management." (For more information about "new forestry" as it evolved in the late 1980s and early 1990s, see Franklin 1989; Salwasser 1990; Sample 1991.) It is a very promising development in forestry. Ecologist and historian Nancy Langston, for example, ends *Forest Dreams, Forest Nightmares: The Paradox of Old Growth in the Inland West* with a chapter aptly entitled, "Living with Complexity," in which she concludes that the path to better forestry lies in "giving up our ideals of maximum efficiency and commodity production, and substituting other ideals which allow for complexity, diversity, and uncertainty" (Langston 1995, 306).

Past experience suggests that resource management outcomes can be only partially controlled (if at all) over the long term because of the incredibly complicated and shifting physical, social, political, institutional, and economic environments. However, past experience also indicates that certain patterns in resource management outcomes repeat over long periods, despite changes in physical conditions, social contexts, and even managers' intentions. I lump these consistent tendencies into two general points: National forest resources are used primarily to implement economic policy, and the Forest Service does essentially what it is paid to do.

Regarding the first point, managers, policymakers, and budgeters have always emphasized activities that produce or enhance marketable commodities, whether timber, forage, water, fish, game, or scenery. Despite strong support for "nonmarket" values (for example, preserving biological diversity, protecting nongame wildlife, reducing soil erosion, restoring degraded ecosystems, and preserving wilderness) and a great deal of recent talk about "ecosystem management," past experience suggests that activities that generate wealth or encourage market exchanges probably will continue to dominate the Forest Service budget and management orientation. Until very recently, the prevailing tendency was to manage the national forests for the highest feasible yields of commercial goods and services while maintaining minimum standards of environmental protection. Decades of battles over endangered species, wilderness designations, the road and trail systems, overgrazing, and silvicultural practices all provide ample evidence. (This theme runs throughout Hirt 1994 [especially chapters 9–12].) Although recent national forest policy pronouncements—especially under Chiefs Thomas and Dombeck—have emphasized forest condition rather than forest products and services as a management goal, I predict that the larger institutional tendency to focus on goods and services will continue to override policy rhetoric. After all, concern over forest condition gains political currency essentially because healthy forests presumably support the highest sustained yields of goods and services.

Those who favor returning to the high–timber harvest levels of the 1960s through the 1980s might argue that since 1990, this emphasis on commodities over environmental protection is no longer the case because endangered species preservation appears to have overridden timber production objectives. But I argue that the crash in timber harvests is due to much more complicated factors than simple environmental protection mandates. Timber harvests have substantially declined on almost all national forests over the past decade, not only those with endangered species conflicts. Other contributing factors include a decline in national forest timber supply, opposition to below-cost timber sales, market competition, more balanced interdisciplinary input into forest management decisions, and improved information provided by the forest-planning process (Hirt 1994, chapter 12).

Regarding the decline in timber supply, as I show in *A Conspiracy of Optimism* (Hirt 1994), the Forest Service permitted and even promoted extensive overcutting of the national forests from the 1950s to the 1980s, rapidly liquidating most of the high-volume old growth (as Marion Clawson [1975] strongly advocated in the 1970s), to allow rapid conversion to younger, faster growing timber plantations. This hopeful yet flawed policy sought to achieve maximum timber growth rates for maximum future productivity, but it led to a traditional timber boom and bust—ironically, the very thing Congress created the Forest Service to guard against. The postwar national forest timber program was unsustainable and enormously damaging to soil, water, and wildlife resources. The timber program today is not dead—only diminished and better integrated with other management objectives. Logging levels today are comparable to logging levels of the immediate post-World War II period, before the Forest Service decided to liquidate its old growth and "maximize" timber production. The timber maximization effort was a failure and a public relations disaster. Fortunately, current harvest levels are "sustainable" now. Perhaps the painfully downsized, dependent mills and rural communities can expect a steadier, even if lessened, long-term supply of logs based on a more conservative, selective harvest of trees that also protects resources while allowing modest use.

My second point—that the Forest Service does what it is paid to do—is an incredibly important and obvious, though largely overlooked, fact. As I mentioned in my discussion of the Coronado National Forest and as amply demonstrated by various scholars already cited, it makes little difference what the agency says it wants to do if Congress fails to provide funding to accomplish those objectives. The annual battles over timber harvest targets in the House and Senate Appropriations Committees during the 1980s are obvious examples of this direct connection between funding and management. The best way to reform forest management, then, is to reform the Forest Service budget.

Scholars such as Randal O'Toole (1988), in contrast, convincingly argue that the best way to reform the Forest Service is to reform its incentive structure. I think he is correct, in part. The agency's budget is indeed a powerful source of incentives that, if altered, would reshape forest management. But O'Toole suggests that government-based budgetary incentives (appropriations) are fundamentally flawed and irredeemably corruptible, whereas market-based incentives are rational and desirable. He would eliminate most federal subsidies and make the Forest Service earn its budget by selling resources at prices people are willing to pay. He predicts that this approach would reduce destructive resource exploitation (by eliminating logging road subsidies, for example) and increase the preservation of environmental amenities because people are willing to pay for scenic views, wildlife, campgrounds, and wilderness experiences.

O'Toole and other "free-market environmentalists" may be partly correct, but history provides plenty of evidence that markets are just as flawed an incentive structure as political appropriations when it comes to achieving environmental stewardship goals. Debilitating environmental deterioration, economic instability, and social irresponsibility resulted from unregulated market transactions on private forest lands in the late nineteenth century, and it led to the conservation movement and the founding of the national forests in the first place. Neither government nor markets offer a panacea. I would no more wish to have all lands and resources privatized than I would have all lands and resources socialized. A cooperative regime of market incentives and government management is a worthy compromise between the opportunities and the pitfalls of either one alone.

Of course, other factors besides congressional appropriations and market incentives significantly influence management, such as professional bias, public values, and resource conditions. But if we want to understand past forest management tendencies and future management trajectories, then we must consider the strongly controlling influence of the federal budget process. And, as I noted in my first point, policymakers and budgeters tend to emphasize marketable commodities (timber, forage, water, fish, game, and tourism). "Ecosystem management" and biodiversity protection are likely to be low on the priority scale, even if they are prominent in the policy debate—which brings us to a vision of the future.

WHERE WILL THE FOREST SERVICE BE IN 2010?

Where I would like to see the Forest Service ten years from now is not necessarily where I expect the Forest Service to be. Looking back over a hundred years of federal forest management tends to sober one's predictions about the future. I begin with my sense of where the Forest Service will likely be in ten years and follow with a summary of where I would *like* to see the agency.

The Forest Service, as an agent of government policy, will continue to be driven by appropriations (budgets) designed to promote economic and social objectives. As usual, these objectives will represent a mix of economic welfare policies, keyed to the market, combined with social policies, keyed to public values, which may or may not involve marketable commodities. I have argued that the relative balance between the production of commodities and the preservation of amenities has shifted over time and according to circumstances: Before World War II, when demand for national forest resources was low, management was predominantly "custodial," even though the Forest Service wished to be more actively involved in timber management. As demand for these resources increased during and after the war, the scales tipped markedly toward commodity production. This imbalance occurred because the opportunity for economic development existed and because the federal government favored economic growth policies.

In recent decades, as national forest timber supplies have dwindled and production costs escalated, the scales have tipped back toward a balance between production and preservation. This improved balance in management exists, I believe, mainly because the great majority of the valuable, marketable timber on the national forests has been captured and liquidated. Old growth and wilderness are now so scarce as to drive up their value in the public mind, whereas the economic, social, and environmental costs of continued extensive clearcutting has hampered efforts to sustain high harvest levels. A rhetorical commitment to multiple use, sustained yield, and forest protection did not ward off an exploitative thirty-year timber boom and bust after World War II that was fostered by market opportunities and pro-development government policies. Now—and for the next few decades at least—changed conditions and opportunities provide us a second chance to get it right, albeit with fewer resources and much greater environmental restoration needs. Unfortunately, the Forest Service's growing interest in restoration will likely be frustrated by the federal government's larger tendency to skimp on funding management activities that have no direct economic benefit.

During this transition to a third era in national forest management, we can expect the specific mix and proportion of forest uses to change in response to markets, prices, public demands, and the availability of resources. Timber simply will not be available in the volumes and quality and at the low production costs that existed during the booming postwar decades. Old growth will remain a very small part of the timber harvest base for a long time. Smaller trees, used for pulp and chips rather than for dimension lumber, will dominate. Importantly, we also will continue to see new justifications for the timber harvest program: Since the early 1990s, the Forest Service has talked less and less about sustained yield (because the agency failed to sustain the yield) and more and more about salvage logging and silvicultural treatments for forest health. (See the series of news articles, speeches, and reports on forest health at the USDA Forest Service Web site, http://www.fs.fed.us/foresthealth/.) This new rhetoric promises to be with us for a long time because it better reflects current forest conditions, timber supply limitations, political opportunities, and changing public values.

I remain skeptical, however, that the change in language reflects a genuine shift in institutional orientation. Timber management will almost certainly remain a "harvest" program that responds fundamentally to market opportunities and pressure from mills. In the past, salvage logging often has served as a justification for harvesting timber in areas where the costs of logging far outweigh the value of the trees being removed (Hirt 1994, 113–7, 165–6). "Salvage" implies an unusual or emergency situation in which something (timber values) needs to be saved. The rhetoric of emergency becomes important because timber salvage operations usually engender huge financial losses to the Forest Service; "emergency salvage" then becomes the only possible justification for logging in such instances. However, we would save money and probably reduce overall biological harm by leaving the trees—even dead and dying trees—in the forest to recycle into soil. Salvage logging and "forest health treatments" have thus become the predominant new defense for timber harvesting, especially where financial losses are particularly high. I expect this trend to continue, and I think the public should be wary of it.

Because timber sales and related revenue likely will remain well below the boom levels of the 1950s through the 1980s for at least the next several decades, investments in timber-related projects and infrastructure probably will not increase much either. Most important, such investments would include timber access road funds. It is no surprise that Chief Dombeck successfully implemented the first nationwide road construction moratorium for roadless areas of the national forests just a few years after the crash in national forest timber production (Forest Service News 1998). Other formerly timber-related management activities such as reforestation, soil erosion control, and wildlife habitat restoration also are likely to suffer over the long term. However, as with funding for endangered species and historic resources, institutional support for these environmental restoration activities probably will rise and fall erratically and unpredictably in response to political changes and temporary policy initiatives (for example, the erratic shifts in funding in response to policy initiatives that I discuss in Hirt 1994, chapters 9 and 10).

In contrast to the reduced timber supply and consequent falling timber management budgets, recreation resources probably will continue to grow in popularity and demand—a trend that began in the 1950s—leading to increased organizational and budgetary support for the management of recreation resources. But as recreation becomes increasingly commercialized and high-use recreation areas suffer more damage to soil, water, and vegetation, I believe the government will treat recreation less as a democratic prerogative (as it was viewed in the postwar decades) and more as a consumptive use of the forest subject to escalating regulation, user fees, and professional administration.

As the U.S. population grows, the national forest resource base correspondingly shrinks, in relative terms, which inevitably leads to more competition for those resources and more conflict. At the same time, the growth and diversification of organized groups interested in the national forests means that the Forest Service's constituency will grow increasingly complex. Because an agency's identity and priorities are partly determined by its relationships with its constituents, the Forest Service is likely to continue suffering from internal squabbles that reflect the competing interests of its constituencies and the growing diversity within the ranks of the agency itself. Rather than recapturing a Golden Age of organizational cohesion attributed to earlier eras in the agency's history, complexity and diversity are likely to reign, stymieing efforts to construct any singular identity or mission for the Forest Service in the next century. I suspect that ten years from now, public controversy over resource allocations and problems with agency morale will be as prevalent as at present, although the Forest Service may be more adept at handling competing demands and internal conflicts if it learns from current experience. History suggests that the agency can be as resistant to social and political change as it is adaptable (resistance to change is the main thesis in Clary 1986).

Two changes that may help the agency cope with conflict are clearer policy mandates and more consistent funding. With vague or conflicting mandates and erratic year-by-year funding, neither the agency nor its constituents know what to expect or how to move systematically toward management objectives. Something else that may help in this new era of increasing conflict and uncertainty is an institutional capacity to adapt. Once again, conservative management that maximizes options, minimizes risks, provides safety buffers for unexpected contingencies, and avoids irreversible commitments of resources offers the best chance to adapt to change and meet public expectations for responsible land stewardship.

An interesting trend developing over the past decade or so is an increasing tendency toward interagency communication and cooperation as well as more collaboration between governments and nongovernmental organizations. Federal (and state) agencies have a long and glorious history of insularity, balkanization, and bitter turf battles—best exemplified by the famous antagonism between the Park Service and the Forest Service, which lasted more than 80 years (for more information, see Rothman 1997), and the scandalous rivalry between the Army Corps of Engineers and the Bureau of Reclamation (for more information, see Reisner 1993). For various reasons, including stifling bureaucratic complexity, litigation, and declining fiscal resources, recent administrations have succeeded in encouraging key players to sit down at the table together to negotiate conflicts and coordinate resource management efforts. This is an interesting and promising development, because negotiated decisions are more likely to have institutional support and coordinated management offers the opportunity to pool resources and expertise.

Associated with this trend is some evidence that boundaries (physical and organizational) are becoming a little more permeable, authorities more shared, and management more collaborative. The evidence for this trend is mainly anecdotal, but it suggests possibilities for cooperation and conflict resolution. Examples include some of the many instances in which Forest Service managers—with institutional support—are cooperating in watershed or regional multi-agency planning efforts (for example, the Interior Columbia Basin Ecosystem Management Project), working more closely with landowners and municipalities in the "urban interface," supporting collaborative conflict resolution efforts such as the Quincy Library Group, and accelerating discussions on land exchanges and consolidations. I believe these opportunities will increase in the future along with the need to exploit them.

WHERE *SHOULD* THE FOREST SERVICE BE IN 2010?

The first and most important institutional reform, and probably the most difficult, should be to increase the *implementability* of the forest plans: make the forest plans more realistic, then provide the institutional support to implement them. A forest plan based on untenable assumptions or unlikely funding levels is little more than a letter to Santa Claus. A reasonable, implementable plan without adequate, consistent funding is not worth the paper it is written on. This initiative would advance the unfulfilled intentions of RPA and NFMA.

Specifically, reforms should include new forest-planning regulations that require planners to (1) disclose historic line item budget trends in the Forest Plan EISs; (2) constrain proposed actions to realistic budget levels; (3) include contingency plans for shortfalls in funding, that is, make specific commitments regarding which activities will have priority if projected funding levels are not realized; and (4) make accomplishments of resource restoration objectives a *prerequisite* to continuing resource development when the two activities are related—for example, reforestation and logging, range improvement and deferred animal unit month reductions, and so forth. Most of all, we somehow need to get a firmer commitment from Congress to provide balanced, integrated funding that enables balanced, integrated management.

Second, the Forest Service should harness the power of the market when practicable to meet overall resource management goals. But markets are double-edged swords that should be utilized cautiously. As always, there will be instances in which the forces of supply and demand should be resisted or constrained to meet conservation goals. "Marketizing" resource allocation may improve management in some areas but will likely hinder management in others. The Forest Service should respond to and use markets but not enslave itself to them.

Revenue from fees charged to forest users must continue to be supplemented by direct appropriations to facilitate comprehensive, integrated, and balanced long-term management. The long-running debate over below-cost timber sales, low grazing fees, and other forms of user subsidies is a complicated issue without an easy answer. Rather than taking the radical position of ending all subsidies, I would like to see the Forest Service and Congress move toward ending subsidies for "consumptive uses" of forest resources—make economic activities pay for themselves—while maintaining subsidies for custodial, restorative, and research activities. Admittedly, there is a gray area between consumptive and nonconsumptive uses (many dispersed forms of recreation, for example), and many "restoration" needs may in fact arise as the external effects of economic development, making appropriated funds for restoration a subsidy for development. Thus, the issue is cloudy and certain to be contentious. But at least we have the privilege to debate and negotiate these questions in the public arena.

Third, I would like to see the Forest Service maintain a better balance between its production functions and its stewardship functions. Historically, the agency has spent a great deal of rhetorical energy on its natural resource stewardship responsibilities, but since at least World War II, its physical energies have been almost wholly dedicated to serving public demands for resource commodities. Although this focus has changed somewhat recently, old habits die hard. As a public land managing agency dedicated to forest and water conservation, the Forest Service always should constrain economic development of the forests to the larger goal of conserving the soil and water that provide the foundation for healthy productive forests. Responding to market demands for resources is certainly part of the mission of the Forest Service, but another central mission of the agency has been to resist purely market forces in pursuit of a broader range of both market and nonmarket goals such as watershed protection, sustainable development, and multiple use. I hope the Forest Service can actually implement the vision articulated by Chief Dombeck in "Natural Resource Agenda for the 21st Century" (summarized in *The Pinchot Letter* 1998, 11 and 12), in which watershed restoration, sustainable forest management, and learning "to live within the limits of the land" are his top priorities. But, as I mentioned earlier, words on paper do not have the power to determine the future. We will need broad-based commitment, public support, legal authority, and organizational resources to implement this vision.

Finally, I would like to see the Forest Service take greater advantage of ecological restoration initiatives. Restoration is appearing commonly as a management objective in many developed areas as well as remote areas of the national forests that have suffered the effects of fire suppression, destructive logging, overgrazing, or insect and disease outbreaks. Theoretically, ecological restoration is central to the Forest Service's mission; however, institutional support for restoration has been lacking, leaving nature largely to its own healing devices. This laissez-faire approach to environmental rehabilitation has proved distressingly inadequate. The Forest Service faces a daunting array of problems with soil erosion, reforestation failures, water pollution, and loss of native biological diversity. The need for restoration is compelling, and the time is ripe. Perhaps this new era of fewer resources (biological as well as organizational) will stimulate greater commitment to sustainable land stewardship, which entails a continuing effort to restore areas to health and productivity during and after extractive uses. Policy initiatives wax and wane in popularity. I see ecological restoration on the rise and therefore an opportunity for the Forest Service to take a lead in this emerging field.

Along with this opportunity comes a danger, however. "Restoration" has many meanings and many more management implications. A wide array of interests will struggle over the definition and application of restoration objectives: Environmentalists will seek to restore "natural" conditions, ecological diversity, and wildness to areas in need of restoration, and the timber and ranching industries will advocate intensive management to restore the economic productivity of degraded areas. These two opposing approaches to restoration are reflected in the recent "forest health crisis" debate in the Northwest, especially in the context of the salvage-logging initiatives of the last decade. One side seeks to close roads and minimize human intervention in natural processes while the other side seeks to build new roads and intensify management activities—both in the name of restoration. The extractive industries especially have become adept at promoting timber sales under the guise of restoring ecosystems. The Forest Service must thread its way carefully among these political land mines, balancing the two approaches.

The debate over restoration reflects a larger conflict at the foundation of contemporary forest management controversies. The path to conflict resolution also leads toward my own desired vision for the future of the Forest Service. The conflict is between the environmentalist's fundamental distrust of forest management as it has been traditionally practiced and the fundamental faith in forestry expressed by the timber industry and the Forest Service. From my point of view, the historical record on both private and public forests provides more justification for the environmentalist's skepticism than for the forester's confidence. Past management failures are the primary source of that distrust. And I believe those failures, in turn, result largely from the post-World War II abandonment of a conservative approach to land management and the failure to equitably balance competing values and uses. The phrase most commonly used by the Forest Service to describe its postwar approach to forestry was "intensive management for maximum production" (Hirt 1994, chapter 6 [especially 141–4]). From the 1940s through the 1980s, the Forest Service proved far too willing to abandon margins of safety and gloss over environmental damages in pursuit of "full use and development" of national forest resources.[1] This management approach maximized the production of commodities while, at best, maintaining minimum standards of environmental protection. This professional orientation, exacerbated by unbalanced funding from Congress, is largely what got the Forest Service into trouble in recent decades. Rejection of this paradigm offers the best route for the Forest Service to avoid management failures and the resulting public disappointments in the future.

The replacement paradigm, based on a commitment to ecological restoration, would re-enshrine humility over hubris; conservative management over intensive management; and healthy, resilient forest conditions over sustained commodity outputs. Such a paradigm would give the agency some elbow room to better adapt to change. It would provide the flexibility to accommodate a greater diversity of management objectives, but not a greater quantity. And it probably would help to heal some of the public mistrust and internal dissension that currently plagues the Forest Service.

I am fully aware that for a service-providing organization, the most unpopular and politically dangerous course of action is to voluntarily reduce the level of services it provides to its constituents. No federal agency is likely to do this on its own. The Forest Service needs public support and institutional support. Americans need to lower their expectations of what the national forests can provide in terms of goods and services as they raise their expectations regarding quality land stewardship. Congress, in turn, needs to accept and support the agency's recent downsizing of the timber program and its efforts to shift to an ecosystem management paradigm.

The 1990s have been a decade of crisis in national forest management, and in 1998 we were in the throes of a major watershed change. The timing of Resources for the Future's conference on the future of the Forest Service was propitious in that the contemporary crisis offered an opportunity for change that was not available even ten years prior. In many ways, the radical environmental movement that proliferated in the 1980s forced an important but painful public debate about environmental values and priorities, and the judicial system in the 1990s helped end Forest Service "business as usual" with endangered species lawsuits, timber sale appeals, and so forth. These two direct interventions, along with increasing congressional micromanagement of the agency, slowed down the inertia of past traditions and practices enough to precipitate a crisis among those dependent on the status quo, thus opening opportunities for reform.

National forest management today is very different from national forest management fifteen years ago. Both the direction and the mission of the agency are in question. The next decade will set important precedents. In our efforts to construct a new mission and a new management paradigm for the Forest Service, we should carefully ground our expectations in historical understanding so that our judgments and prescriptions are as realistic and implementable as possible.

ENDNOTE

[1]See, for example, the precedent-setting Forest Service document *Full Use and Development of Montana's Timber Resources*, published by the USFS Northern Regional Office, Missoula, MT, in 1958 [Senate Document 9, 86th Cong., 1st Sess., Jan. 27, 1959]. This document was touted as a model to be emulated by other regions of the National Forest System.

REFERENCES

Clarke, Jeanne N., and Daniel McCool. 1985. *Staking Out the Terrain: Power Differentials among Natural Resource Management Agencies.* Albany, NY: State University of New York Press, 38.

Clary, David A. 1986. *Timber and the Forest Service.* Lawrence: University Press of Kansas.

Clawson, Marion. 1975. *Forests for Whom and for What?* Baltimore, MD: Johns Hopkins University Press for Resources for the Future, 30.

Coronado National Forest. 1993. *Forest Plan Five Year Review Report,* FY 1987–FY 1991. Tucson, AZ: Coronado National Forest Supervisor's Office, 5.

Culhane, Paul J. 1981. *Public Lands Politics: Interest Group Influence on the Forest Service and the Bureau of Land Management.* Baltimore, MD: Johns Hopkins University Press.

Dietrich, William. 1992. *The Final Forest: The Battle for the Last Great Trees of the Pacific Northwest.* New York: Simon and Schuster, 254–64.

Egan, Timothy. 1990. *The Good Rain: Across Time and Terrain in the Pacific Northwest.* New York: Vintage Books, chapter 9.

Farnham, Timothy J. 1995. Forest Service Budget Requests and Appropriations: What Do Analyses of Trends Reveal? *Policy Studies Journal* 23(2): 253–67.

Farnham, Timothy, Cameron Taylor, and Will Callaway. 1995a. A Shift in Values: Non-Commodity Resource Management and the Forest Service. *Policy Studies Journal* 23(2): 281–95.

———. 1995b. Racial, Gender, and Professional Diversification in the Forest Service from 1983 to 1992. *Policy Studies Journal* 23(2): 296–309.

Forest Service News. 1998. *Roadless Areas and New Transportation Policy.* http://sv0505.r5.fs.fed.us:80/forestmanagement/html/roadlessnews012198.html.

Franklin, Jerry. 1989. Toward a New Forestry. *American Forests* 95(11/12): 37–44.

Hirt, Paul W. 1994. *A Conspiracy of Optimism: Management of the National Forests since World War Two.* Lincoln, NE: University of Nebraska Press.

———. Forthcoming. Institutional Failure in the U.S. Forest Service: A Historical Perspective. *Research in Social Problems and Public Policy.*

Kaufman, Herbert. 1960. *The Forest Ranger: A Study in Administrative Behavior.* Washington, DC: Resources for the Future.

Langston, Nancy. 1995. *Forest Dreams, Forest Nightmares: The Paradox of Old Growth in the Inland West.* Seattle: University of Washington Press.

LeMaster, Dennis C. 1984. *Decade of Change: The Remaking of Forest Service Statutory Authority during the 1970s.* Westport, CT: Greenwood Press, chapter 2.

O'Toole, Randal. 1988. *Reforming the Forest Service.* Washington, DC: Island Press.

Reisner, Marc. 1993. *Cadillac Desert,* revised ed. New York: Penguin Books, chapter 6.

Rothman, Hal. 1997. A Regular Ding-Dong Fight: The Dynamics of Park Service-Forest Service Controversy During the 1920s and 1930s. In *American Forests: Nature, Culture, and Politics,* edited by Char Miller. Lawrence: University Press of Kansas, 109–24.

Salwasser, Hal. 1990. Gaining Perspective: Forestry for the Future. *Journal of Forestry* 88(11).

Sample, V. Alaric. 1990. *The Impact of the Federal Budget Process on National Forest Planning.* New York: Greenwood Press.

———. 1991. *Land Stewardship in the Next Era of Conservation.* Milford, PA: Grey Towers Press.

Steen, Harold K. 1977. *The U.S. Forest Service: A History.* Seattle: University of Washington Press, chapter 10.

Symposium on Change in the U.S. Department of Agriculture Forest Service. 1995. *Policy Studies Journal* 23(2): 245–371.

The Pinchot Letter. 1998. Washington, DC: Pinchot Institute for Conservation, Spring issue.

USDA Forest Service. USDA Forest Service Home Page. http://www.fs.fed.us/foresthealth/.

West, Terry L. 1992. *Centennial Mini-Histories of the Forest Service.* FS-518. Washington, DC: USDA Forest Service, 69–72.

Wilkinson, Charles F., and H. Michael Anderson. 1985. Land and Resource Planning in the National Forests. *Oregon Law Review* 64(1/2): 138–55.

Discussion

Predicting the Future by Understanding the Past

Mark Rey

Trying to predict the future of the Forest Service based on an understanding of the past underscores the most fundamental problem inherent in making public policy in this area. Everyone—or almost everyone—takes license to create his or her own version of the past to validate his or her desired future. When I meet new reporters who are covering this beat, I try to brief them on some of the "rules of the road" for what is a very complicated area of public policy. One of these rules is that everybody, including me, is entitled to his or her own opinion and own set of facts. The sooner you figure that out, the easier it will be to file your stories.

I find it as difficult to find evidence of a "conspiracy" in the Forest Service of the past as I do finding any "optimism" in the Forest Service of today. But let me express the "conspiracy of optimism" in another way. That isn't exactly how Paul Hirt expresses it, perhaps, but it is nevertheless the way I've heard it from other practitioners of the theory. The abbreviated version of the past is that the Forest Service (1) optimistically projected levels of outputs that were neither ecologically sustainable nor, in many instances, socially desirable; (2) offered these projections to secure progressively larger budgets; and (3) used these progressively larger budgets—and I almost quote from Chapter 7—to maximize output production while, at best, maintaining minimum standards of environmental protection; which in turn (4) left the national forests with—and now I directly quote Paul—"a daunting array of problems with soil ero-

MARK REY is on the staff of the U.S. Senate Committee on Energy and Natural Resources.

sion, reforestation failures, water pollution, and a loss of needed biological diversity."

I have trouble finding data to fully support or, in some cases, even partially support any of these four indictments. So, let's take each one in turn.

UNSUSTAINABLE OUTPUTS

Here, we are talking mostly about timber and less about grazing. First of all, with respect to timber, it was in the late 1960s that the Forest Service made the calculations that suggested that continuing to ramp up timber sales and subsequent harvest levels would not result in "sustainability" over the long term, as sustainability was defined at that time. Levels were constant from the late 1960s until the late 1980s. Most important, the levels were viewed as sustainable based on what was known, believed, or hypothesized at that time.

Since then, we have learned more, we have redefined the term, and we have adapted to and adopted other values and more data. Our definition of sustainability today is different from that used in the late 1960s, even the late 1980s.

I don't think we can say that the agency—based on what was known at the time—provided outputs that were unsustainable.

INFLATED BUDGETS

First, the simple fact is that inflated budgets were never even requested. In 1976, then-Chief John McGuire testified before Congress during the consideration of legislation that was to be the National Forest Management Act (NFMA) of 1976, immediately after the agency produced its first long-range program in 1975: the Resources Planning Act (RPA) Program that was required by 1974 RPA legislation. That testimony should be where you would find evidence of the conspiracy, if there was one. Indeed, that testimony should be the equivalent of the Zapruder film for conspiracy theorists. If you can't find the man on the grassy knoll in that testimony, I would suggest that he's probably not there.

At that time, McGuire was in possession of the first long-range plan and called in front of a relatively desperate Congress, given the events of the Monongahela decision. So, he was in a position, it seems to me, to instruct them to pay up in small, unmarked bills and to leave him alone afterwards. But what he said was,

> We believe the recommended program is a good start in planning today for the needs of tomorrow. Decisions on initiating

> federal actions to achieve the general program goals must, however, remain consistent with overall decisions on the size of the federal budget and other budget priorities. We recognize that we are not now able to implement the program goals at the preferred level. These are not propitious times in which to initiate long-term, relatively expensive programs; however, we believe that the path we have charted is in the right direction and our fiscal resources, at whatever level they may be, can be better directed as a result of the assessment.

This passage (McGuire 1975) seems to me a pretty lackluster call for inflated budgets. Indeed, it appears to be a recognition that planning for an ideal is very different from securing the funding necessary to achieve that ideal. This concept was recognized and reflected on by Congress.

If we are going to accuse the agency of asking for inflated budgets, someone is going to have to show me where and to whom the request was made. I would submit to you that those inflated budgets were never requested. But perhaps more important, they were never granted. Congress never granted anything close to what the RPA Program had suggested. Some budget increases were granted, and they were used to balance out other resource objectives and to secure greater levels of resource protection based on developing information and science at that time. And that's the point of Table 1.

In Table 1, you can see what the Forest Service was actually doing in 1975; what the first plan predicted, with the desired budget, they could

Table 1. The 1975 Resources Planning Act: The Way We Were, Are, and Thought We'd Be

		1996		
	1975 Actual	*Predicted*	*Actual*	*2020 Predicted*
Developed recreation (RVDs)	72.8 million	97 million	145 million	122.1 million
Dispersed recreation (RVDs)	125 million	173 million	195 million	220 million
Wilderness (acres)	12 million	20 million	34.7 million	30 million
Wildlife and fish habitat improvement (acres)	179,000	1.9 million	166,611	2.8 million
Range (AUMs)	11.3 million	16 million	9.2 million	20.4 million
Timber (cubic feet)	2.7 billion	3.5 billion	0.76 billion	4.07 billion

Note: All predicted values are from the 1975 Resources Planning Act. RVDs are recreation visitor days; AUMs are animal unit months.

produce by 2020; and what the Forest Service actually produced in 1996. The agency is producing not only far less than what it predicted it would produce in the mid-1990s but also less than it actually produced in 1975. However the additional money might have been spent, I don't believe it was spent to produce more range, more timber sales, or more commodity outputs.

MAXIMIZING OUTPUTS

I have already suggested that it's hard to find any available data to support the idea that outputs were maximized while environmental protection was minimized. But I can also suggest to you that protection has grown progressively as we have learned what it is we are trying to protect. This effort has been ongoing since the turn of the century, when the Forest Service came into existence.

AN ARRAY OF PROBLEMS

Soil erosion, reforestation failures, water pollution, loss of native biological diversity—the available data don't support the proposition that today's national forests are ecological wastelands. In fact, considerable data—as well as a wealth of personal experience from anybody who has traveled in, recreated in, and otherwise enjoyed the national forests—suggest exactly the opposite. The national forests still are a highly desired natural resource.

The propositions of widespread soil erosion, reforestation failures, and water pollution are relatively difficult to support in light of the available data on the ground. The only open question is in the protection of native diversity. I argue that we are still trying to define what this term means—for both the national forests and, for example, the Tidal Basin in Washington, DC, where the issue is whether we ought to let native beavers come back and eat the Japanese cherry trees that surround it. We are still debating what exactly we are trying to protect, and it seems reasonable that we should resolve that debate and define what it is we want before we accuse the Forest Service of failing to provide it.

LOOKING TO THE FUTURE

After considering the points raised by Paul Hirt, we realize that the question has become, "What does and should the future hold?" In this case, although we agree on almost none of the past, I find myself in unusual—one would probably say, almost bizarre—agreement with Paul. We are in the midst of a period of intense political debate about what the future holds, and that debate will continue to intensify. I also suspect that as our population ages and becomes more sedentary, we will take our recreation from arguing about the national forests, rather than visiting them.

As to what the future should hold, I find myself agreeing with Paul. It is reasonable for the plans to ensure greater implementability. That should happen. The agency should disclose historic line-item budgets in the forest plan environmental impact statements. We should try to constrain proposed actions to realistic budget levels, and include contingency plans for shortfalls in funding. We should make accomplishments of resource rehabilitation objectives a prerequisite for continuing resource development—here, I argue that we do in most cases. However, in each of these areas, Congress will have to provide this direction. I don't believe that any administration is going to willingly allow the Forest Service to disclose this sort of information as part of the annual budget process, because it will diminish the administration's own ability to set larger priorities in how federal budgets are constructed. This is where the disconnect has occurred. Congress has not failed to provide the funds requested by the Forest Service; rather, those funding levels were never transmitted to Congress as part of the Forest Service budget request because the administration had far more pressing priorities in every given year.

Which brings us to Paul's last recommendation, that most of all, we somehow need to get a firmer commitment from Congress to provide balanced, integrated funding that enables balanced, integrated management. Left unstated is the implied desire for Congress to provide more funding to the Forest Service to achieve that balance; this is where Paul and I part company. It is not reasonable for me, as a representative for members of Congress, to suggest that more funding will be forthcoming. In fact, it won't. It's time to get over it and get used to it. We still are operating in an era of budget deficits; the budget surplus being debated in Congress is an illusion. It is a surplus created by the Social Security Trust Fund, which itself will be bankrupt in short order unless we find some way to deal with it. The notion that we are going to provide an agency that controls the kinds of assets that the Forest Service has responsibility for managing with additional funding is something that I don't think is going to happen—not in this Congress, not in past Congresses, not in future Congresses. That leaves us to find ways to use these assets—both commodity and noncommodity—to generate revenues and to make investments that are sound. Some ideas are as simple as the administration's proposals in the 1998 budget—to charge reasonable fees for commercial photography on the national forests and reinvest those fees—or, at the other end of the spectrum, to look to taking some of these lands off the budget and create other management systems, as Randal O'Toole (1988) and others have suggested.

But if you think the solution is to troop down to Capitol Hill en masse and say the only thing we all can agree on is that the Forest Service needs more money, and Congress should cough it up, then I think I can save you the trip.

REFERENCES

McGuire, John. 1975. Congressional Testimony.

O'Toole, Randal. 1988. *Reforming the Forest Service.* Washington, DC: Island Press.

8

Does the Forest Service Have a Future?

A Thought-Provoking View

Roger A. Sedjo

The Forest Service is in transition. Change is everywhere. The agency knows where it has been but has a much less clear vision of where it is going. Just before leaving the Forest Service in late 1996, then-Chief Jack Ward Thomas stated, "The Forest Service needs a revision, or at least, a clarified mission" (Thomas 1997, 182).

Historically, the Forest Service mission has been fairly well defined. The 1897 Organic Act gave three purposes to the forest reserves: (1) to preserve and protect the forest within the reservation, (2) to secure favorable conditions of water flows, and (3) to furnish a continuous supply of timber for the use and necessities of the people of the United States. In the intervening years, the mandate was gradually expanded to include other of the forest's multiple uses. The most recent comprehensive forest legislation, the National Forest Management Act (NFMA) of 1976, mandates the Forest Service to "provide for multiple use and sustained yield of the products and services obtained therefrom ... and, in particular include coordination of outdoor recreation, range, timber, watershed, wildlife and fish, and wilderness."

The legislation appears unequivocal: The Forest Service is to provide for the sustainable production of the seven products and services explicitly mentioned. The outputs are clearly identified, as is the requirement

---

ROGER A. SEDJO is a senior fellow at Resources for the Future (RFF) and director of RFF's Forest Economics and Policy Program.

that they be produced on a sustainable basis. If this legislation is so clear, then what is the rationale for Thomas' lament?

Thomas went on to explain some of his concerns: "It is not yet widely recognized—much less openly acknowledged—but public land managers now have one overriding objective (or constraint) for management—the preservation of biodiversity." He gave his view of the inadequacy of the legislative support for these activities: "The law does not clearly say [the Forest Service should manage for the preservation of biodiversity]. Nobody seems to openly recognize it." Additionally, he stated, "I don't personally have an objection to [managing for the preservation of biodiversity]—if that is what society wants. The Congress and the President need to examine the situation that has evolved and ask, 'is that what we intend?' If so, so be it. If not, then clarification is required as to what is expected of federal land management agencies in regard to achieving 'multiple-use' management" (Thomas 1997, 161).

Many of these activities have been driven by the Endangered Species Act (ESA), which requires the protection of habitat for threatened or endangered species as well as other environmental legislation. The courts and recent federal policy held that the requirements of the ESA are clear and overriding. If conflicts occur between ESA and an agency's other governmental statutes, ESA must dominate. The stress on biological considerations, however, has been made even stronger than that of the ESA by the "viability clause" in the Forest Service regulations. Developed to support the NFMA, these regulations require the Forest Service to ensure the widespread maintenance of viable populations of plants and animals.[1] Many recent court decisions hinge on the viability clause.

The result of these conflicting signals, Thomas suggested, is that in recent years, there has been a serious disconnect between the directives of the Forest Service's statutory mandate and the nature of the activities and management being practiced by the Forest Service. This disconnect is due, in no small part, to a host of intervening litigation and court rulings, and Thomas believes that clarification is required by Congress and the administration. In fact, no legislative clarification has been forthcoming in the period since Thomas stepped down as chief in late 1996. The most recent authorization statutory legislation is still the NFMA of 1976, which calls for management for sustainable production of a set of multiple outputs, whereas the de facto practice of the Forest Service, according to Thomas (1997), has been to manage for the preservation of biodiversity.

Recently, the Secretary of Agriculture assembled a Committee of Scientists (COS) to "provide scientific and technical advice to the Secretary of Agriculture and the Chief of the Forest Service on improvement that can be made in the NFS [National Forest System] Land and Resource planning process." In its report (USDA 1999), the committee decided to provide the new mission statement that the Forest Service has lacked. Casting aside concerns about whether it was appropriate for the committee to dictate a mission for the Forest Service, the committee boldly declared that the binding charge has been sustainability and recommended, in essence, that the Forest Service manage for sustainability. Apparently, the committee was less concerned about the necessity of having a legislative directive to provide mission clarification from Congress and the President than was Chief Thomas. Furthermore, an articulation of what ought to be the focus of management clearly is not a scientific question but a reflection of personal values. Thus, in addressing the issue of what "ought" to be the objective of management, the committee went well beyond what its scientific credentials could justify. In fact, some members of the committee asserted that the manager's obligation to provide for species viability and ecological integrity is "morally" appropriate.

Having asserted a mission for the Forest Service that Congress and the administration were reluctant to state, the committee then suggested ways that this objective might be accomplished. The COS report (USDA 1999) argues that sustainability is paramount and, in essence, that the legislative multiple-use mandate ought to be replaced de facto with this alternative objective: maintaining ecological sustainability.

One can argue the appropriateness of COS in establishing social objectives and attempting to tie personal preferences to morality. However, the fact that the COS recommendations are at such variance with the Forest Service's statutory multiple-use mandate highlights the environment in which the Forest Service has been forced to operate and the potential contentiousness that might occur in the absence of an effective, well-defined mission. In today's society, land-use decisions have become "moral" issues, even among presumably objective scientists.

In this chapter, I examine the past and current situation of the Forest Service and try to provide a contemporary perspective. First, I briefly cover the history of the Forest Service, most of which is well known. Next, I describe and characterize the recent and current situation in which the Forest Service finds itself, with a discussion of the major problems and challenges. Finally, I outline a number of alternative possible future scenarios for the Forest Service, highlighting some of their strengths and weaknesses.

BACKGROUND

In response to public concern over water conditions and future timber supplies in the latter part of the nineteenth century, large areas of public lands were designated as part of the nation's "forest reserves," later to be called the National Forest System (NFS). However, even then, people held alternative perspectives and philosophies regarding the objectives of forest maintenance. The pragmatism of the conservationists, as represented by Pinchot, was reflected in their concept of the "wise use" of resources. The philosophy of wise resource use was pitted against the views of preservationists, such as Muir and perhaps Thoreau. Both views had followings. The American people wanted water and timber, but they also were concerned about preserving naturalness, wildness, and wilderness—which even then were recognized as part of the American heritage.

Although these two philosophies struggled for dominance over the years, the on-the-ground conflicts between these two perspectives were small—largely because the forest land managers assumed primarily a custodial role. The public forest provided only modest amounts of timber, allowing the vast majority of public forest to remain largely unaffected. The wise-use philosophy prevailed in the early twentieth century, as Pinchot and his conservationist successors dominated the institution of the Forest Service.

Using Clawson's characterizations (1983), one can perhaps view the first fifty years of the Forest Service history, from its inception to about World War II, as a period of custodial management and forest protection; however, it also was active in rehabilitation in some locations. During and after World War II, the national forests took on a new importance as a source of timber. They produced substantial amounts of timber, first to meet the needs of the war and then for the post-war housing boom; in fact, high levels of output would continue into the late 1980s.

As the 1950s gave way to the 1960s, however, public environmental concerns were growing. Among these concerns was a fear that the NFS emphasis on timber was too great and that focus should also include other forest outputs. The Multiple Use Sustained Yield Act of 1960 emphasized nontimber goods and service outputs provided by the forest as well as the sustainability of these outputs. Production of multiple outputs, however, generated concerns over the level and mix of the various outputs. As the rancor among the various interests grew, Congress passed the Resources Planning Act (RPA) in 1974 and the NFMA in 1976.

The RPA, particularly as amended by the NFMA, was crafted to address the source of the contentiousness. The NFMA legislation tried to do at least three things. First, it tried to articulate a multiple-use vision. Specifically, this vision called for the production of multiple outputs, including timber, range, wildlife, recreation, water, and (less explicitly) wilderness. The "trick" was to produce these outputs jointly and to produce the mix that would satisfy the various publics. In addition, the laws required that these outputs be produced in a sustainable manner.

Given this general mandate, a forest-planning process was created that was intended to allow all of the interested parties to participate in management and output decisions. The assumption was that the planning process would provide a vehicle for the various interests to work out their differences and converge on a consensus forest plan with a broadly acceptable mix of actions and outputs. Also, it was implicitly assumed that if consensus were reached on the forest plan (that is, the goals of forest management in a particular forest), then Congress would provide the budget to implement those objectives. In addition, the importance of monitoring these resources on a continuing basis had earlier been addressed by a provision in the 1974 act for the periodic renewable resources assessment.

In the more than two decades since the NFMA, little of what was envisioned has become reality. Although periodic resource assessment has been undertaken, the planning process can, in many respects, be viewed as a failure. For example, it has not generated the desired consensus. In the first 125 forest management plans were about 1,200 appeals and more than 100 subsequent lawsuits. Some appeals have been in process for almost a decade without resolution.

Furthermore, when plans were created, budgets generally were not forthcoming to allow faithful implementation. Little or no relationship has developed between most of the plans and the budget. In fact, two largely independent planning processes now exist: one, the "forest planning process" is called for in the NFMA and involves protracted "public participation" by the various interested "publics"; the other is that undertaken by the administration and Congress in their deliberations regarding the budget to be provided to the Forest Service. There is little connection between the budget that emerges from the political process to provide funds on an aggregate programmatic basis and the various forest plans developed through the decentralized planning process (Sample 1990).

NO LONGER AN ELITE AGENCY

Traditionally, the Forest Service had been viewed as an elite agency. This perspective emerged out of the ties between Pinchot and President Teddy Roosevelt and the prevailing progressive philosophy (see Nelson 1998) that placed confidence in technocratic solutions. This was a new agency with a new mandate supported by the President. Gifford Pinchot, who later became governor of Pennsylvania, had the power of the President behind him and was able to craft an agency relatively insulated from the usual bureaucratic and political pressures commonly directed at agencies such as the Forest Service. The new agency would reside not in the Department of the Interior, which was viewed as highly political, but in the Department of Agriculture. There, the agency could have both a high degree of power and maximum autonomy.

This organizational location reflected the Forest Service responsibility for not only protection but also the active promotion of tree growing, restoration, and research. Consistent with the positive view of progressivism and scientific management, the Forest Service was able to recruit the best and the brightest foresters trained in new European techniques. This new agency sported a highly trained and committed professional staff. The view of professionalism was maintained for many years. Until the early 1990s, the chief of the Forest Service was still essentially a nonpolitical position, held by someone drawn from the ranks of the agency's senior professionals.

The Forest Service made the most of its positive image. As recently as the early 1950s, the agency was commended in an article in *Newsweek* (1952) for its professionalism, effectiveness in dealing with forest concerns, and ability to work with local people to help achieve local objectives. The Forest Service had a storehouse of goodwill both in Congress and in the hinterlands where it operated.

However, as a member of the COS, my experience suggests that little goodwill currently exists, especially in the various local regions. The committee held almost a dozen meetings in various regions of the country. One of the most memorable and perhaps most pitiful observations I made from those meetings was the high degree of frustration and disillusionment on the part of local forest users. It was local people—those who raised issues of local use of timber, forage, and summer home permits—who felt the most betrayed by the Forest Service and the process. They had believed that by participating in the process, they could contribute to the final outcome. Ultimately, they found that their hard-fought positions and compromises meant nothing, because the forest plan frequently was tied up in appeals and litigation. Or, perhaps worse, they found that the plan and compromises they worked so hard to achieve were never implemented for lack of budget or because of an overriding executive decision.

In contrast, our COS meetings were regularly attended by—and we received substantial numbers of comments from—representatives of national timber and environmental interests. Often, the same people met with us each week in a different region of the country as they monitored the process while providing both visibility and comment. These people, usually on a payroll, recognized how the political game is played and knew not to take these processes (the COS and earlier planning meetings) too seriously. Nor were they inclined to count to heavily on any promises being kept.

In Kaufman's famous *The Forest Ranger* (1960), the Forest Service was used as an example of how a large public government organization ought to function. He argued that, unique among large organizations, the Forest Service had been able to maintain its focus, discipline, and esprit de corps. The high esteem in which the Forest Service was held was not limited to the public; it carried over to Congress. The Forest Service's rapport with Congress was rewarded by generous budgets. In return, the Forest Service fostered a "can do" attitude. It provided Congress with what Congress wanted. In fact, some experts believe that this very can-do attitude led the Forest Service to press its harvest levels to the limit in order to meet the high harvest desires of its champions in Congress. Ultimately, however, these high harvests appear to have backfired, resulting in increasing numbers of citizens who became concerned that harvest levels were "too high" and an erosion of the balance that had been achieved in its constituencies.

In *Public Land Politics* (Culhane 1980), Paul Culhane argued that the Forest Service had successfully been able to maintain a high degree of autonomy as the various interest groups competed against each other. The groups he examined—timber interests, environmentalists, and recreationists—all provided the Forest Service with constituencies that supported its budget requests and programs. In return, the Forest Service provided the outputs desired by each group. Because the interests were so diverse but relatively evenly balanced, the Forest Service had autonomy in decisionmaking in that it could justify an action undesired by one of the groups by arguing that it was necessary to pacify one of its other constituencies, who wanted even more. Furthermore, when the time for budget decisions arrived, these groups still could be relied on to support the various facets of the Forest Service budget with Congress.

Today, few would view the Forest Service as an elite agency. Local users of national forest lands are highly disenchanted and discouraged. Recreationists, environmentalists, range users, and timber users also voice major complaints. It seems that nobody is happy with the Forest Service.

A quintessential example of the general disillusionment is the experience of the Quincy Library Group, a small, informal group that met in the library in Quincy, California, to discuss issues relating to the management of the several national forests in the region. This group, having given up on "the process," has undertaken direct political action with what appears to be great success so far. Bypassing the Forest Service entirely, it has appealed directly to the California delegation in Congress for a separate management charter and separate funding. Legislation to this end recently passed. With the help of the new legislation, the Quincy Library Group hopes to attain two goals: greater control over what is done on local Forest Service lands and a larger budget with which to manage these lands, according to local desires and objectives. The ultimate success of this endeavor remains to be determined.

THE CURRENT SITUATION

As I suggested earlier, the happy situation of the early Forest Service has seriously eroded over recent decades. My own hypothesis is that the system has broken down because the fine balance among the various competing constituencies gradually disappeared. Battles among these groups—particularly the environmentalists and timber interests, together with the recreationists—compelled Congress to pass the NFMA to try to restore order and balance. However, this was not to be. In addition to increasing rancor over the management of the NFS, a host of environmental laws and their evolving interpretation by the courts forced both a reduction in harvest levels and a rethinking of policy. Timber harvest levels, which peaked in the late 1980s under the still-existing NFMA legislation, have declined since to less than one-quarter of their peak levels.

Whatever its past "sins," the Forest Service has truly been given a "mission impossible" in recent decades. It is being asked to reflect the will of the people when in fact the people in this country are deeply divided. We share no vision of the role of public forest lands, even though the recent American Forest Congress attempted to define a shared perspective for forestry (Bentley and Langhein 1996). Attempts to "reinvent" the role of the Forest Service continue to be frustrated by a lack of consensus. Apparently, according to the picture painted by former Chief Thomas, the Forest Service has reinvented itself.

With the national forests now providing only about 3% of the U.S. industrial wood supply, the nation obtains wood from other sources to meet demand. Harvests on private lands have filled some of the gap, as private owners increase their management intensity. Also, timber imports have risen from foreign suppliers, especially Canada, and today would be even higher were it not for the recent imposition of trade restrictions on certain wood products.

The culture of the Forest Service has changed from the inside, as staff trained in traditional forestry has been supplemented with those trained in wildlife, ecology, and the biological sciences. Although this kind of change may be inevitable and indeed desirable, it also contributes to confusion regarding the appropriate mission for the agency.

Changes also have taken place at the top. Historically, Forest Service career professionals have followed a civil service career path, which allowed them to rise through the ranks to the top position of chief. In recent years, however, the nature of the selection process for the chief has changed. The days that the highest civil service appointment in the U.S. government was the chief of the Forest Service have passed, and the senior positions are now more political in nature.[2] Today, contacts between the professional civil service staff and their politically appointed leaders—including the chief—are more limited than in earlier eras, and many of the agency's resource managers feel that their professional decisionmaking is "micromanaged" by political appointees.

In many respects, the Forest Service is probably more politically vulnerable today than at any time before in its history. The public's former trust in scientific management, which was a major driving force in the creation of the Forest Service under Pinchot and Teddy Roosevelt, is highly eroded, if not in total disarray. The Forest Service also is essentially naked to various political forces because of its lack of any serious constituency.

Last year, several groups came to the defense of the NFS when Republican congressmen fired a shot across the Forest Service's bow by raising the prospect of returning to custodial management (Murkowski and others 1998). If the Forest Service cannot provide outputs for constituents, they argued, then why should we spend large amounts of resources on management? It became clear that the groups that came to the defense of the NFS were defending not the Forest Service but the federal forests. Furthermore, some have opined that if the proposal had come from the Democrats, rather than the Republicans, it probably would have been more warmly received.

The Forest Service has lost many of its traditional supporters. For example, although the timber industry has been a traditional Forest Service constituency, in recent congressional hearings, the American Forest and Paper Association supported only a very modest budget for the Forest Service, noting that it viewed the recent activities and outputs of the Forest Service as of only limited interest to the members of the association. In addition, the industry has been stressing the need for salvage logging to remedy forest health problems found in the excessive timber and under-story buildup, which pose various disease, pest, and, especially fire hazards. Remedial management would provide some timber for local operations.

Environmental organizations that might have been expected to support the new policy direction at the Forest Service have not appeared. National environmental organizations are not interested in active forest management. The split between national and local environmentalists over the nature of the desired management of public lands in the region of the Quincy Library Group reveals deep divisions, even within the environmental community, regarding appropriate forest management.

Similarly, recreational interests are largely absent in defending the Forest Service. Finally, with few exceptions, local forest users appear to be largely frustrated and disillusioned with the Forest Service, and asking a local member of Congress to support overall Forest Service budgets that are seen as only minimally—if at all—responsive to local concerns likely would spark little interest. None of these events bodes well for the future of the agency.

Thus, the Forest Service now stands largely exposed, without public constituencies willing to advocate its cause. Given this lack of power, it is difficult to see how the Forest Service could resist an attempt—such as was made under the Carter administration in the late 1970s—to reorganize the agency, perhaps out of existence. It is doubtful that the Forest Service could find champions to defend it, as it has done so successfully in the past. Indeed, given the absence of a mandate that has broad support, one might ask whether there are any reasons to try to maintain a Forest Service separate from other federal land management agencies.

WHERE FROM HERE?

Clearly, if it is to move beyond its current malaise, operate efficiently, and even survive, the Forest Service needs a well-defined mission and a powerful political constituency. But what do the American people want from their national forests? A host of things.

It probably is accurate to say that Americans want naturalness and an element of wildness. They surely want many of the ecological services provided by forests, including watershed protection, erosion control, and wildlife habitats. The American people might even support a program that identifies the primary responsibility of the NFS as maintaining a sustainable ecosystem and shifts the responsibility for timber production to private producers and foreign lands. Let me examine three potential candidates for a Forest Service mission and constituency: biological preservation, recreation, and local management.

Within the COS was the strong belief that the national forests ought to be managed for biological preservation and ecological sustainability. Furthermore, another objective of managing the forests was to keep the forest condition within the "historic range of variability" that predated Europeans and maintain the objective of returning them to pre-European forest processes and functions. Managing for this objective clearly involves a "mission shift" away from a focus on multiple outputs and would involve dramatically reducing certain outputs from levels of the recent past. These reductions would affect timber as well as other outputs, such as certain kinds of recreation.

In my view, such an approach is, in effect, an obituary for the Forest Service as we know it. It is doubtful that, in the absence of significant tangible outputs, there is sufficient public support to generate serious budgets for a program focused primarily on maintaining ecological sustainability. Although many people may support such an approach in concept, this support probably could not bring together a constituency with the power to generate substantial and continuing budgets for these management activities. The services generated by the activities would be difficult for the public to perceive on a regular basis, and the major direct financial beneficiaries would be the biologists and ecologists employed in the process. Although major environmental groups support facets of an ecological mission, many of the major national groups oppose timber harvesting of any type, including that necessary to meet other objectives (for example, wildlife habitat). Indeed, many favor an essentially hands-off approach to "management."

A "hands off" approach today may result in long-term fundamental changes in the nature of the forest because of the absence of traditional predators, changes in natural disturbances because of human intervention in the recent past, and other changes (such as the influx of exotic species) that have occurred. Because of their persistent distrust of the motives of the Forest Service and separate objectives of their own, many national environmental groups would not enthusiastically support the large budget necessary for the active management for ecological sustainability. More likely outcomes of a focus on management for ecological sustainability would be the erosion of the Forest Service budget and the substitution of benign neglect for active management.

Perhaps the major constituency that could emerge to take leadership in supporting the Forest Service and its management is that of the recreationists. The NFS provides various kinds of outdoor recreation. Even though these groups are far from monolithic in their interests and services desired from the Forest Service, their numbers are substantial. Perhaps most intriguing is the possibility of generating a significant portion of the budget for the forests from recreational user fees. Certainly, many forests have the potential to raise funding from such fees, as some forests located near urban centers currently demonstrate. Such fees often are difficult and costly to collect. Nevertheless, it has been argued that for many national forests, the recreational benefits far exceed the timber and other traditional output benefits. If this is true, then user fees could provide major revenue sources for many, but surely not all, national forests. In this context, Forest Service budgets could be substantially financed out of recreational receipts and supplemented by more modest allocations from Congress. Of course, such an approach would require that the Forest Service have some control over the user fees it generates.

If funding were dependent on recreational use, the Forest Service would have a powerful incentive to provide the kinds of outputs that recreationists desire. Furthermore, the role of federal funding and the ability of a constituency to support the Forest Service budget in Congress become less important if the Forest Service covers a large portion of its costs with user fees. Finally, recreational uses may be in conflict or lead to conflicts with other desired outputs and services, such as biodiversity. Thus, whereas this approach appears to have many commendable characteristics, we are not guaranteed that future conflicts between the various user groups could be avoided.

A third option would be to move in the direction of more localized input into the management of the national forests in different localities, in the spirit of the Quincy Library Group. Perhaps Congress should consider budgeting national forests or groups of national forests individually, by using a manner akin to the budgeting of the national parks. This method could allow management to be customized—to a degree not seen previously—to the needs and desires of the local communities. Some combination of user fees and customized management could provide for both adequate funding and the emergence of powerful local constituencies, which could allow for local participation in a way not experienced in decades. However, many national environmental groups are opposed to this approach. Shifting power to the local community implies reducing the influence of national groups on local situations. Perhaps, more fundamentally, the issue relates to the appropriate division of decisionmaking power over a local resource between the local community and citizens living in other sections of the country. Even though this kind of solution offers promise in that it addresses the budget and constituency challenges facing the Forest Service in a way other approaches do not, it seems unlikely that all forests can expect the financial support likely to be received by the Quincy Library Group.

Perhaps the most fundamental question is whether to retain a separate Forest Service at all. Arguments for the coordination of land management are louder than ever. This recommendation recurred in the COS report (USDA 1999), which suggested improved coordination with not only private forest land but also the various federal agencies. To date, the coordination among federal agencies leaves much to be desired.

The arguments at the beginning of the twentieth century called for a Forest Service focused the creation of an elite organization that had technocratic prowess and a degree of independence from the bureaucratic and political processes so that it could "do the right thing" based on its professional judgment. A primary argument that has been used in the past to fight reorganization efforts is that the Forest Service is more akin to an agricultural agency, focused on producing crops of trees, providing protection, and so forth. Clearly, the production issue is less relevant now. On the cusp of the twentieth and twenty-first centuries—given the recent behavioral objectives of the Forest Service—these motivations are largely absent.

The Forest Service can no longer claim to be an elite organization. Although the agency has retained many highly trained and competent people, it is no longer unique and probably is more wracked with confusion than most agencies because its mission has lacked clarity (or has been highly ambiguous) for many years. Neither is it any longer insulated from the ravages of the bureaucratic process or from crass politics. In fact, former Chief Thomas (1997) stated that "the entire process is becoming increasingly politicized through orders which originate above the chief's level," and where the "exact source of those instructions is sometimes not clear." The fine balance among constituencies, which Culhane (1980) saw as the core of the Forest Service's ability to fend off crass political pressures, no longer exists. Furthermore, its ability, or willingness, to supply services to various constituencies is minimal. It now is beholden to a single group in society rather than a host of groups.

Finally, we may have a compelling reason to integrate federal land management agencies. Although agencies have been directed to coordinate management, many people believe that coordination is inherently more difficult across organizations than within an organization. It has been argued that cross-agency coordination of federal lands has been grossly inadequate. Perhaps it is time to reconsider the proposals of the Carter administration, two decades ago, which called for a unified department of natural resources that would include the Forest Service. Perhaps it is time to merge the Forest Service and the Bureau of Land Management into a single agency. Surely, the rationale for such integration becomes more compelling as the agency loses both its unique mission and its unique ability to perform any mission in an outstanding manner.

FINAL THOUGHTS

The material I have presented here is intended to be provocative, of thought and discussion about what we want and how we can get it—innovative ways to meet current challenges and smart planning to create the future that we envision. Perhaps it is time to "think the unthinkable." The Forest Service has been an unusually successful organization for much of its history, but that is no longer the case. Today, the agency finds itself with legislation that gives it a multiple-use statutory mandate while being covered by the single-purpose ESA statute. This problem is exacerbated by the lack of a public consensus. Until this deadlock is broken, the Forest Service will be in the limbo that Thomas described (1997).

However, if the Forest Service is converted into a biological reserve, it may no longer be politically viable as a separate institution. At a minimum, it is clearly time to rethink the role and mission of the Forest Service. A doable mission needs to reflect the views of a cross section of Americans rather than the values of a single interest group or a small group with a unique set of values. The American people need to enter into a major dialogue, and Congress and the administration must provide clear direction. Furthermore, this dialogue should be expanded to seriously consider whether the federal land management problems of the twenty-first century might require the creation of new streamlined integrated organizations to replace outmoded agencies.

ENDNOTES

[1]The term "viability" does not appear in the 1976 law, which made passing reference to maintaining biological diversity. The viability provision was added and broadened in the process of developing the regulations that support the law.

[2]Historically, the chief was appointed from among the small group of long-term Forest Service employees who were qualified as Career Senior Executives. However, in recent years, chiefs have been either not qualified as Career Senior Executives or not drawn from the ranks of long-term Forest Service employees.

REFERENCES

Bentley, W. R., and W. D. Langhein. 1996. *Seventh American Forest Congress: Final Report.* New Haven, CT: Yale School of Forestry and Environmental Studies.

Clawson, Marion. 1983. *The Federal Lands Revisited.* Baltimore, MD: Johns Hopkins University Press.

Culhane, Paul. 1980. *Public Lands Politics.* Baltimore, MD: Johns Hopkins University Press.

Kaufman, Herbert. 1960. *The Forest Ranger.* Baltimore, MD: Johns Hopkins University Press for Resources for the Future.

Murkowski, Frank H., and others. 1998. Letter to Mike Dombeck, Chief of the USDA Forest Service, from members of the U.S. Senate Committee on Energy and Natural Resources, Washington, DC, February 20.

Nelson, Robert H. 1998. *Rethinking Scientific Management.* Discussion Paper 99-07. Washington, DC: Resources for the Future, November.

Newsweek. 1952. June 2.

Sample, V. A. 1990. *The Impact of the Federal Budget Process on National Forest Planning.* New York: Greenwood Press.

Thomas, Jack Ward. 1997. Challenges to Achieving Sustainable Forests: In NFMA. In *Proceedings of a National Conference on The National Forest Management Act in a Changing Society 1976–1996*, edited by K. N. Johnson and M. A. Shannon. Draft review. December 10.

USDA (U.S. Department of Agriculture) Committee of Scientists. 1999. *Sustaining the People's Lands: Recommendations for Stewardship of the National Forests and Grasslands into the Next Century.* Washington, DC: USDA.

Discussion

Does the Forest Service Have a Future?

R. Max Peterson

Roger Sedjo wrote that he intended for Chapter 8 to be provocative. It certainly is! It also provides a great deal of food for thought, whether you agree or disagree with his conclusions. It also provides a brief, but I believe useful, overview of some of the history of the Forest Service, including the establishment of the forest reserves, which predated the Forest Service.

WHAT ELSE IS NEW?

I agree with Sedjo's assessments that the Forest Service is in transition, change is everywhere, and the Forest Service knows where it has been but has a less clear vision of where it is going. You might say, what else is new? The Forest Service has been changing for the past ninety-four years—changing to reflect society's changing needs and to apply new scientific findings to better manage land and forest resources. It would be both inaccurate and misleading to suggest that the events of the past several years are simply normal and expected changes in an agency. Many significant changes have been made by administrative actions that normally would have been made only after substantial public debate and congressional action.

What is new is the breadth and depth of disagreement between different uses and users and the degree of distrust that permeates current forest policy debates. Disagreement between Congress and the administration also seems to be both more strident and more prominent today

---

R. MAX PETERSON is a former chief of the USDA Forest Service.

than any time in my memory. Certainly, during the time I worked closely with members of both parties in Congress as deputy chief and then as chief of the Forest Service, there were differences of opinion—but not the sometimes bitter exchanges that seem to have become rather common in recent years. I recall a pattern of mutual respect between members of Congress and the Forest Service, even when we had significantly different ideas about policy.

Strongly held policy disagreements are not a new phenomenon. History tells us that even the act of establishing the forest reserves in the 1890s and early 1900s was not universally accepted. In fact, reserving public land as forest reserves was strongly opposed by some people in the forest products and mining industries and was a matter of great concern to many other early users of the public lands. In fact, most of you know that Congress specifically removed the President's authority to reserve land from the public domain for forest reserves. That was done when Teddy Roosevelt was President and Gifford Pinchot was chief. In fact, Pinchot, who many people think was universally popular, was hanged in effigy in such places as Cordova, Alaska, because of his role in preventing land patents from being issued for coal-bearing lands in that state. So sure were people that those patents would be forthcoming that a narrow-gauge railroad was built to the site of the coal reserves! That episode ultimately led to the firing of the first chief.

I mention this little bit of trivia to emphasize that through the years, the public has expected different things from national forest land. Still I admit that to the best of my recollection, I have never witnessed as much disagreement as I have in recent years. Clearly, members of Congress are reflecting what they perceive as strong public disagreement with Forest Service management of national forest lands.

As Sedjo correctly states, the strong historical support for forest reserves came from those concerned about securing "favorable conditions of water flows" and those who had witnessed the cut-over, burned-over, farmed-over devastation of forest land in the eastern United States and feared it would be repeated in the west. They also feared a "timber famine"—hence, support for forest reserves to provide a "continuous supply of timber."

Whether we agree or disagree with the changes in the Forest Service during the past five to ten years, we probably all would agree that they have been rather dramatic and have had significant social and economic impact, particularly on traditional users in the western United States. This effect has caused substantial loss of agency support, particularly by substantial segments of the public and members of Congress. Of course, some who perceive themselves as "winners" support the changes; however, many think the changes are inadequate. Sedjo identified the result as a loss of historical constituencies without replacing them with new ones. I reluctantly agree with that assessment.

Another striking fact is that these changes have been made administratively—not only without changes in laws but also over substantial objections from members and leaders in Congress. In fact, for the first time in my memory—which spans more than half of the lifetime of the Forest Service—a number of traditional constituencies and powerful members of Congress openly suggest that the Forest Service is no longer worth keeping as an agency. For example, Senator Ted Stevens (R-AK), who chairs the full Senate Appropriations Committee; Senator Frank Murkowski (R-AK), who chairs the Senate Energy and Natural Resources Committee; and Congressman Don Young (R-AK), who chairs the House Resources Committee, recently were quoted as agreeing that "there is no reason for your being" (quoted in *Anchorage Times* 1999). Unfortunately, the same sentiment has been expressed by other prominent members of Congress.

Does this mean the Forest Service needs a new mission, as suggested by Chief Jack Ward Thomas (1997)? I disagree. I don't believe that such action is feasible or helpful. Also, the likelihood that Congress and the administration would agree on a mission statement that would be helpful in resolving current polarization over management and use of national forest land is somewhere between slim and none. Several years ago, I resolved not to spend a great deal of time heading up blind alleys, flying up box canyons in airplanes, or trying to make changes in laws that obviously are not feasible!

Rather than a new congressionally mandated mission, I believe what is needed is a new consensus or at least a workable majority of the public deciding that responsible *sharing* of national forest resources for this and future generations is better than the current, rather dysfunctional situation. There is also a need to find public processes that encourage sharing to replace the current rewards for pursuing political decisionmaking, litigation, and appeals. If that happens, Congress and the administration could again trust the Forest Service to make scientifically sound, professional decisions.

As long as organized interest groups see their interests better served by political influence, litigation, or appeals, we likely will see pendulum swinging, increasing polarization and disenchantment with Forest Service programs and policies. That situation could lead to a level of disenchantment that ends the Forest Service as an agency. I doubt that will happen, but I must admit I am less sure of that prediction now than I was even a year ago.

A third practical problem for the Forest Service not mentioned by Roger Sedjo is the loss of trusted, experienced leadership at national, regional, and local levels. Coupled with increased decisionmaking by political appointees above the Forest Service, this means that many people consider decisions by the Forest Service as both unimportant and meaningless because they probably will be reversed. Forest Service people in the field are thus less likely to make tough, controversial decisions.

DOES THE FOREST SERVICE NEED A NEW MISSION?

First, in my view, there are four major assignments that the Forest Service has by law and policy; none has been expressly stated in statute as missions. Generally stated in laws as programs, activities, or purposes, they include the following:

- National and international leadership in forestry.
- Forestry research to serve both public and private landowners.
- Cooperation with the states to further improve management of private forest land.
- Management of the 191 million acres of national forest lands. Current national forest statutory direction clearly provides that "the Forest Service should manage to provide for multiple use and sustained yield of the products and services obtained therefrom and in particular include coordination of outdoor recreation, range, timber, watershed, wildlife and fish, and wilderness" (the National Forest Management Act [NRMA] of 1976).

Thus there are statutory *purposes and uses* to be served by national forest, but no statutory mission statement per se. The statutory direction for the Bureau of Land Management (BLM) is similar. It includes statutory purposes but not a mission statement.

Similarly, until 1997, there was no statutory mission statement for National Wildlife Refuges. The one that was adopted is, "The mission of the System is to administer a national network of lands and waters for the conservation, management, and where appropriate, restoration of the fish, wildlife, and plant resources and their habitats within the United States for the benefit of present and future generations of Americans" (National Wildlife Refuge System Improvement Act 1997, Sec. 4). Interestingly, there was no organic act for the National Wildlife Refuge System until that law was enacted in 1997.

Without quibbling over semantics, let me point out that a mission statement is supposed to broadly state the reason an agency exists. For example, the commonly stated mission of the U.S. Department of Defense is to "provide for the common defense" of the United States, which is reflected in the preamble of the U.S. Constitution. In theory, after mission or purposes come goals, uses, and more clearly defined objectives to accomplish those goals and accommodate the uses to fulfill the purposes of the organization. All of these elements must be within the framework and the specific mandates of enabling legislation and policy.

PURPOSES OF NATIONAL FOREST MANAGEMENT

Sedjo's chapter relates entirely to the management of the national forests—not the multifaceted purposes or programs of the Forest Service. From this point on, I focus on management of the national forests, what the purposes of management have been, and what I believe the future purposes should be.

Public lands are governed by myriad laws—some of which overlap and sometimes contradict objectives of other laws. It would be helpful to the Forest Service and BLM if the administration and Congress would agree to objectively look at how a multipurpose agency can function effectively when a number of single-purpose federal agencies can dictate management actions or stop action on national forests based on a judgment or opinion of that agency. The Constitution of the United States gives Congress the responsibility "to make all laws which shall be necessary and proper." The most specific congressional statement about the purpose of the national forests is contained in the NFMA of 1976, which I quoted earlier.

The purposes to be served by national forests have changed since 1897, when the Organic Act gave the three purposes for the forest reserves, which Roger Sedjo outlined: (1) to preserve and protect the forest within the reservation, (2) to secure favorable conditions of water flow, and (3) to furnish a continuous supply of timber for the use and necessities of the people of the United States. Clearly, in 1897, the primary reason that forest reserves were created was that public forest and brush-covered, primarily mountainous, unoccupied land in the west were being cut over, mined, overgrazed, and otherwise exploited to the point where erosion, fire, and lack of management would spell long-term devastation of that land as well as private land in valleys downstream. In 1911, the Weeks Act recognized similar purposes for eastern national forests. Forest reserves had strong support from western states such as Colorado and California, which had experienced the destruction of irrigation systems from silt and floods. Private ownership of forest land in the east, which most of the population related to, resulted in cut-over, burned-over, and frequently abandoned land. Thus, forest reserves were seen as an answer to secure a favorable condition of water flow and provide a continuous source of timber. For several reasons, the management of the forest reserves was transferred to the U.S. Department of Agriculture, and the Forest Service was established in 1905. Gifford Pinchot and Theodore Roosevelt were driving forces behind this change. Part of the reason for this transfer was to place management of federal forest land in the department that already had responsibility for both providing assistance to private landowners and researching natural resources.

I suggest that the conceptual mission or purpose for the forest reserves as forcefully espoused by Gifford Pinchot and Teddy Roosevelt—that is, to protect, manage, and enhance those public lands so they can provide the goods and services that constitute the greatest good for the greatest number in the long run—is still valid today. Obviously, that conceptual mission is ever-changing and must be reflected in more specific purposes, programs, and plans. Such is normally the case for mission statements.

As he stated in *Breaking New Ground,* Pinchot (1998) recognized that everything was connected to everything else, and that management of the forests must consider a wide range and different mix of goods and services from time to time and from place to place, all within the ability of the forest to sustain those uses. In fact, Pinchot's idea that forests could be perpetuated over time and used conservatively to produce a mix of goods and services was rather radical at the time. Most people thought that a cut-over piece of forest land was worthless, and many acres of such land were abandoned and sold for back taxes.

When wild places began to disappear, people such as Marshall, Carhart, Leopold, and others helped articulate and evolve the concept and reality of wilderness—a place untraveled by humans, where humans are visitors and do not remain (according to the Wilderness Act of 1964).

Recognizing that the emphasis on various purposes changes from time to time is clearly shown in the activities of the 1920s, '30s, '40s, and '50s; national forests provided much-needed employment in the 1930s, a strategic source of timber and minerals during the 1940s war years; lumber for houses for returning veterans in the 1950s; and substantial wilderness acres in the 1960s, '70s, and '80s. National forests, particularly in the east, played a major role in reestablishing wildlife populations that had been decimated before 1940. Although beginning in 1924, when the Gila wilderness was established, wilderness received substantial new impetus by the passage of the Wilderness Act of 1964. Beginning in the 1930s and accelerating after World War II, national forests became widely used and prized by people for outdoor recreation, including fishing, hunting, camping, picnicking, skiing, and boating. Today, national forests provide about 40% of the federal visitor use—more than any other agency.

By the 1960s, the mission per se had not changed in my view, but the mix and level of uses specifically recognized the changing public needs and desires through the passage of not only the 1960 Multiple Use Sustained Yield Act but also the 1964 Wilderness Act and such major acts as the National Trails Act, the Wild and Scenic Rivers Act, and the Endangered Species Act. In fact, controversy over how to manage national forests continued in the 1960s and early 1970s, leading to the famous Monongahela court decision, the Resources Planning Act of 1974, and the NFMA of 1976. Senator Hubert Humphrey, one of the major authors of the NFMA, said the purpose of the act was to provide direction to the Forest Service, so that the Forest Service could return to practicing forestry in the woods and not in the courts. Unfortunately, that was not to be. What would have happened without those acts, though, is open to speculation.

Let me quickly note that the NFMA includes only a half-dozen main ideas:

- There would be one plan per national forest or combination of national forests.
- Interdisciplinary analysis and planning, which included an evaluation of options, would be done as part of the planning process.
- The public would be involved in planning. Thus, a National Environmental Policy Act (NEPA)-type process was embedded in the act itself.
- An orderly method of amending, changing, or revising a plan would ensure relative stability and no surprises to the public.
- Several specific mandates related to such issues as protection of soil and water values, diversity and viability of plant and animal species, and a specific reaffirmation of the Multiple Use Sustained Yield Act of 1960 and the Wilderness Act of 1964.
- A few words were removed from the Organic Act of 1897, which had restricted the Forest Service to selling only large old-growth trees.

As Roger Sedjo pointed out, several arguments and polarization related to national forests are ongoing and severe. I suggest that they are related to the level and mix of goods, services, and benefits that national forests are expected to provide as well as how and to what extent some fairly new concepts of biological diversity or ecological stability are either implicit or necessary so that national forests might meet the expectations of the American public over time.

Underlying these arguments is the question of which of the potential customers of national forests should be served first—or, which use is more important, or what level and type of uses should be provided. I know firsthand the violent arguments that can occur between user groups that are in disagreement over how a national forest or a portion or combination of forests is to be used. Each group may feel quite righteously that its use is the most important one and that other uses should be either eliminated or at least reduced to the point where there is no interference with their more important use. Ironically, some of these strident disagreements occur between the same general kind of users, such as float boaters versus jet boaters on the water, horseback users versus hikers on the trails. The most longstanding arguments though are the so-called commodity versus noncommodity uses of national forests. Some users try to demonize or completely banish other uses, such as grazing, timber harvesting, or mining.

Unfortunately, political involvement at the assistant secretary, undersecretary, secretary, Office of Management and Budget (OMB), the Council on Environmental Quality, and Justice Department levels during the past twenty years has tended to exacerbate those conflicts. The increasing number and frequency of political appointees throughout the natural resource agencies of government—whether in a Democratic or Republican administration—tends to cause significant pendulum swings in the management of public lands. I have witnessed those pendulum swings up close and personal! Without a doubt, they are worse today than ever before.

During the past twenty years, I have been one of those to suggest that a bipartisan commission would be a better answer to getting both professional and more stable management of natural resources that would be most beneficial to the public over time. By their very nature, forests represent long-term public investments and resources; they do not lend themselves well to pendulum swinging or politically motivated management, which tends to operate on a two- to four-year time cycle.

I also believe that a two- to five-year budget and a process that shows budgets by national forest *might* help provide more public ownership and involvement. It also would better link the Forest Resource Management Plan and the budget process. To many people, including a substantial number of Forest Service employees, the current budget process is an annual time-consuming and rather mysterious exercise.

Roger Sedjo argues rather forcefully that the Forest Service is no longer an elite agency with high esprit de corps and that the balance among the various competing constituencies for the ear of the Forest Service and the administration has gradually disappeared. Sadly and somewhat reluctantly, I have to agree with that assessment. I was hopeful that the recent American Forest Congress, which tried to develop a shared vision of the role of public forest lands, would help bring about a consensus. Unfortunately, that was not to be the case, and I think the blame falls on both ends of the spectrum, where it seems that many groups would rather "fight than switch."

In recent years, we have created a situation where there is not much incentive for groups that are in disagreement over national forest management to sit down together and see how these lands could be *shared* for the mutual benefit of a reasonable mix of users in a way that both would meet current needs and would retain the productivity and capability of the forests to meet future needs. As long as people can use appeals, litigation, and the political process to prevail over other users at very little cost, they will be inclined to do so. It obviously helps in fundraising to show great national struggles and "wins." The Quincy Library Group is just one of the manifestations of a group that became wary and disillusioned with the planning process and decided to take matters into their own hands to come to grips with how an area of national forest land should be managed. I admit to having substantial problems with that approach, because it would delegate to a local group—without being required to provide the funding for or to consider other broad public interests—the responsibility for determining the management of significant lands that are presumed to have regional and national significance. I began my career at Quincy on the Plumas Forest fifty years ago, and I can tell you that a 1949 Quincy Library Group would not have been impressed with either sustainability or ecological diversity. A major traumatic reduction in rate of timber harvest was under way to reduce harvest to long-term sustained yield levels.

I understand that in some situations, this kind of approach may be the most viable or most likely way to break stalemates. I have been more impressed with local watershed groups that have cooperated to solve common problems or voluntary coordinated resource management efforts, which usually include multiple public and private ownership.

WHERE FROM HERE?

In his rhetorical "Where from Here?" section, Roger Sedjo argues for a well-defined mission and a powerful, productive constituency if the Forest Service is to move beyond the current malaise. I agree that the Forest Service and any other agency must have strong public support but doubt that a new mission statement is either feasible or useful in causing that to happen.

A new mission statement would be a tempting idea if the Forest Service operated in an atmosphere where agreement could be expected, but any observer of the recent strong differences of opinion between Congress and the administration would recognize that such a constituency is not likely to appear or to be successful in practice. I also argue that the public lands belong to all of the people of the United States, not only a particular constituency. Each of us is a shareholder of national forest and should not be disenfranchised by more powerful users. This concept is important and means that national forest goods and services for a large and diverse number of publics must be considered, not only the most powerful group in today's politics.

To underscore that idea, if we look back, we note that the Forest Service—particularly national forests—served quite different purposes in the 1930s, 1940s, 1950s, 1960s, and 1970s. You may or may not recall the shortage of timber for house construction and the rapidly escalating price of housing in the late 1970s. This situation caused President Carter to issue an order to sell an additional one billion board feet of timber from the national forests to try to restrain lumber prices so that people could afford houses. Likewise, the shutoff of the oil supply in the early 1970s because of problems in the Middle East caused an immediate search for energy resources on all kinds of lands. The President declared that energy self-sufficiency was the "moral equivalent of war."

I personally have witnessed two periods, the 1930s and the 1960s, when public land including the national forests became important for employment and training of disadvantaged youth and for other people out of work. National forests also can provide substantial demonstration areas for enlightened scientific forest management to private forest landowners and for other countries. In fact, one of the major challenges facing forestry around the world today is the exploitation of forests in many countries that continues without renewal. Even though that practice may seem remote to the United States, there can and have been substantial impacts on the global environment, including wildlife populations and markets from such activities. Increasing worldwide populations and widespread poverty do not bode well for many forests of the world. Most forests today are harvested for fuel or to make way for food production.

Roger Sedjo suggests that combining the Forest Service with one or more other agencies might both improve coordination of land management and resolve some of the conflicts that the Forest Service now finds itself faced with on a daily basis. I posit that the opposite is probably true. Larger agencies tend to be less responsive, more difficult to manage, more politically controlled and directed, and less effective in carrying out their missions. One only has to look at major conglomerations of agencies such as Health, Education, Welfare, and the General Services Administration to see what happens when several organizations are put under one large umbrella. It seems to me that the National Park Service, the National Wildlife Refuge System, and the Forest Service have individual purposes or missions that are different enough that cross-agency coordination is certainly required; however, *combining them into a single agency would be a mistake.*

You can make a conceptual argument that the Forest Service and BLM should be combined because they are both multipurpose agencies, and lands are frequently intermingled. As many of you know, I attempted to develop a major Forest Service–BLM interchange to make more sense out of the current public land patterns but found that it was not politically feasible. I doubt that combining the Forest Service and BLM would lead to improved management, and certainly, the fundamental problem of polarization of the public over the purposes or objectives of management would not be solved by that combination. I think it would simply provide more uncertainty and perhaps simply deflect the real issue of trying to get a better public consensus as to how multipurpose public lands should be managed. Unfortunately, reorganization is often a substitute for progress. Usually, layers of management and cost increase concomitantly.

WRAPPING IT UP

Let me provide some final thoughts on the future, because Roger Sedjo has indeed, and appropriately, rung the alarm bell. I had hoped that the work of the recent Committee of Scientists might produce some processes that could be helpful in bringing people together. I'm saddened to say that I believe the reverse is true. Rather than providing some streamlining and suggestions on how to bring people together in the planning process, in my view, the committee took on a role of determining, if not redefining the purpose or mission of the Forest Service. The committee suggested that certain nonstatutory overriding purposes should be the controlling factors. It also added significant complexity and cost to an already complex process. By adding levels of planning, meaningful public participation is precluded, and planning has become a substitute for action. In other words, the Forest Service was looking for a life vest and was tossed an anchor.

If we looked into a crystal ball, we might see that the future holds continuing change in public expectations for the national forests. It might show something like I have hypothesized here.

Eastern United States

Here, where the national forests are rather small but may provide the bulk of the public land available for public use, public desires and needs will warrant a valued mix of goods and services from those forests. This mix probably will include (1) stable watersheds that provide high-quality water; (2) forests that provide abundant fish and wildlife for a variety of outdoor recreational uses, including hunting, fishing, wildlife photography, and so forth; and (3) some particular outdoor recreation uses relating to dispersed uses of streams, lakes, mountains, and wilderness areas.

These lands could continue to be managed to retain forest conditions so that they will be attractive and provide some high-quality wood products; some personal wood supply; and some highly strategic minerals, including oil resources. These outputs could be achieved within a mix that represents the vision of the Multiple Use Sustained Yield Act of 1960 and is the "harmonious and coordinated management of various resources, each with the other, without impairment of the productivity of the land" (Multiple Use Sustained Yield Act 1960).

Western United States

These national forests are quite diverse, so it is not a simple task to envision future management direction for such lands. The national forests of southern California and parts of the Sierras and the front range in Utah, Colorado, and Arizona, for example, will have high values for watershed protection and for a large variety of outdoor recreation, including hunting, fishing, wildlife observation, camping, skiing, and boating. Some of those forests are capable of providing sustained high-quality and important timber resources, which can be provided without significant impact on other values of the forests. Those forests already have large and highly valuable wilderness areas.

Some western national forests contain lands of very high value for fish and wildlife that are declining and may be approaching "threatened" or "endangered" levels. Frequently these lands, along with other public and private land, will be vital to both recovery of species or preventing them from declining to the point of being threatened or endangered. It is simply inappropriate and unworkable, though, for national forests alone to either prevent most species from becoming threatened or endangered or being adequate for recovery. Other public and private lands can or must play a role, or we will all be frustrated by lack of success. A source of dedicated funds for nonregulatory, incentive-based programs is needed for private landowners who manage their lands to provide important public benefits such as habitats for declining or threatened species. Regulatory programs are primitive and costly, and many times too late.

Some lands also are capable of providing high-quality range for domestic livestock within a carefully managed land stewardship program. Some of those public lands also contain highly strategic minerals, including substantial oil reserves that may be vital to national security and to the future welfare of the United States. There is a clear need to determine better ways to lease and to obtain minerals from these lands, in my view, than the current 1872 Mining Law and the 1920 Mineral Leasing Act provide.

SUGGESTIONS

A few suggestions that might be helpful are worth restating:

- Establish a bipartisan commission to provide policy and program direction for the Forest Service that includes national forests.
- Establish a high standard for courts to review decisions of that commission, similar to the standard for regulatory commissions. Courts are a rather questionable venue to contest resource management plans.
- Conduct a detailed review of current statutes and transfer to the commission the responsibility of complying with federal laws that affect national forests rather than other federal single-purpose agencies.
- Reject the current report of the Committee of Scientists and order a redesign of the planning process to make it simpler, less costly, more user-friendly, and more understandable to the public. Insulate the plans from both political and legal review except in certain limited circumstances.
- Remove the undersecretary position from line authority over Forest Service programs and personnel actions. Provide that the chief and other agency heads report to the secretary. (This was the case until after 1960.)
- Revamp the budget process to reflect funds required to implement Forest Plans and set forth multiple-year budgets, national forest by national forest.
- Make the chief and other natural resource agency head positions term appointments that do not coincide with Presidential elections. A seven-year term is suggested.
- Experiment on a regional, state, and local basis with new processes that promote *sharing* of resources of national forests.
- Examine new ways of reducing the effect of high-impact uses so that a mix of uses will be more acceptable.
- Emphasize national forests as demonstration areas to improve national and international forest management.

To wind up this discussion on a somewhat positive note, it seems to me that the well-known Pogo quote, "We have met the enemy and he is us," is an appropriate view of the current status of management of national forests. These lands are a great legacy. They were bequeathed to us by our forefathers to share, for the benefit of not only this generation but our descendants. Instead of seriously considering how we can share these lands, many groups today seem more interested in fighting than switching. The whole idea of conservation assumes that we will share and be good stewards of today's resources on the public lands, both for this generation and future generations.

Each of us has a figurative share of the national forests in our pocket. Another user of the national forests does not have the right to tell you or me that we cannot have any benefit from our share and that they would like to use their share and ours too. So, the fundamental problem as I see it is not only a new mission on paper for the Forest Service but also the hard work of developing a consensus, or at least a substantial majority, on what are the goods and services that the public of today and tomorrow expect as we share and "care for the land and serve people." Without that new agreement, the future of the Forest Service or any other multipurpose agency is potentially imperiled.

REFERENCES

Anchorage Times. 1999. April 20.

Pinchot, Gifford. 1998. *Breaking New Ground,* Commemorative Edition. Washington, DC: Island Press, 365.

Thomas, Jack Ward. 1997. Challenges to Achieving Sustainable Forests: In NFMA. In *Proceedings of a National Conference on The National Forest Management Act in a Changing Society 1976–1996,* edited by K. N. Johnson and M. A. Shannon. Draft review. December 10.

9

The More Things Change…

The Challenge Continues

V. Alaric Sample

As I look back over the Forest Service's first century, I see that despite inevitable change, many things have remained the same. The Forest Service and the national forests were born out of strife and controversy over the appropriate balance between forest conservation and the development of natural resources. The dialogue continues today in a uniquely American political free-for-all, within a system seemingly designed to invite discourse and debate—often passionate. This situation presents a continual challenge for resource managers and users. Considered from a wider perspective, it also reaffirms the democratic ideals for which the national forests were established and the overarching conservation mission that has guided the Forest Service throughout most of its history.

SOCIAL EQUITY AND THE NATIONAL FORESTS

Compared with other developed nations, America is extraordinarily competitive and entrepreneurial. More than any country in the world today, America and its major institutions seem to embrace Adam Smith's suggestion that the public interest can best be served by the collective actions of individuals acting in their own enlightened self-interest (Smith 1796). As a society, we have consistently rejected the level of government regulation and intervention on behalf of social equity that characterizes most European democracies.

V. Alaric Sample is president of the Pinchot Institute for Conservation in Washington, DC, and of Grey Towers National Historic Landmark in Milford, Pennsylvania.

In relative terms, we have relied much more on the laws of the marketplace. We have freed the natural entrepreneurial spirit and, in so doing, have helped create the richest nation in the world. Perhaps recognizing the inequities that inevitably arise in such a loose and exuberant system, Americans also have the highest rate of individual charitable giving of any country in the world. Nevertheless, a huge gap in wealth distribution remains; as much as 85% of the nation's assets are owned by as few as 15% of its citizens.

At the turn of the century, the difference in income and wealth distribution between the richest Americans and average citizens was even more stark. This situation helped give rise to the progressive movement, of which Gifford Pinchot was a proponent. Progressivism profoundly shaped Pinchot's views regarding forests and forestry in the United States.

In Pinchot's mind, the national forests were established largely to ensure that the forests then in the public domain would be managed to meet the diverse needs of the broader American public rather than for the enrichment of a few individuals and interests who held most of the political and economic power at the time. In this concept, Pinchot reflected the values and ideals of the progressive movement. His highly public jousts with the dominant coal, timber, livestock, and water barons over the use of the public lands were celebrated in countless newspaper editorials and political cartoons.

Pinchot's notoriety was not enough to keep him from being fired by President William Taft for blowing the whistle over a sweetheart deal involving the U.S. Department of the Interior and federal coal lands in Alaska, but his martyrdom on behalf of the public interest helped secure his sainthood and rally even more converts to the cause of conservation.

THE LIMITS OF SCIENTIFIC MANAGEMENT

Pinchot's famous dictum was that the national forests were to be managed "for the greatest good, for the greatest number, for the longest time" (which he himself credited to scientist W. J. McGee, "the brains of the Conservation Movement" [Pinchot 1998]). However, he never precisely specified how the greatest good was to be determined, or by whom. No less a scholar and practitioner of public land management than Marion Clawson, whose memory we honor with this volume, felt it necessary to keep asking these fundamental questions seven decades after the establishment of the Forest Service (Clawson 1975).

As Robert Nelson has pointed out in Chapter 4 and elsewhere, Pinchot and the early Forest Service were enamored of "scientific management" as it applied to forestry, and decisions were largely deferred to trained professionals who acted on the basis of the best available science (Nelson 1995). There is ample evidence, however, that even Gifford Pinchot himself had serious second thoughts about the Forest Service's application of scientific management to the national forests (Miller 1992). Decades after serving as chief of the Forest Service in the first decade of the twentieth century, he found it necessary to act as the "public conscience" of the agency he had helped establish, confronting its leaders whenever their actions seemed to advance special interests over those of the general public or stray from the agency's original conservationist mission. Gifford Bryce Pinchot, the forester's only son, would later adopt his father's critical posture. During the "clearcut crisis" of the 1970s, he invoked his father to rebuke the Forest Service for its focus on timber production at the expense of other public values.

The Forest Service's reliance on models of scientific management and economic efficiency reached its pinnacle—some say its Waterloo—during the early efforts at rational comprehensive planning for national forest management that took place in the early 1980s. These models presumed that our scientific understanding of forests, as a social and economic as well as a biological construct, was essentially complete and that the models were capable of comprehending and quantifying all important values and relationships.

However, a group of the nation's preeminent forest scientists, brought together by the National Research Council (NRC 1990) to evaluate the status of forestry research in the United States, concluded that

> the existing level of knowledge about forests is inadequate to develop sound forest management policies. Current knowledge and patterns of research will not result in sufficiently accurate predictions of the consequences of potentially harmful influences on forests, including forest management practices that lack a sound basis in biological knowledge. This deficiency will reduce our ability to maintain or enhance forest productivity, recreation, and conservation as well as our ability to ameliorate or adapt to changes in the global environment.

In other words, our scientific understanding of forests is neither complete nor sufficient.

The NRC report went on to conclude that, as limited as our understanding of ecological processes and functions of forests might be, our understanding of their social and economic context is even more so. In the end, after enormous investments in the development of these complex computerized decision models designed around economic efficiency criteria, the results proved of little value to resource managers and even less to a skeptical public.

MULTIPLE-USE MANAGEMENT AND THE BALANCE BETWEEN ECOLOGICAL AND POLITICAL OPTIMALITY

In Chapter 5, Binkley describes his vision for the future management of the national forests. He foresees some combination of overarching land use designations such as plantation forests, parklike preserves, and natural forests. This view is essentially the same as the "triad model" of forest land use advanced in recent years by Hunter and Calhoun (1996) and Seymour and Hunter (1992; forthcoming). The triad model reflects in some ways the Forest Service's existing de facto approach to implementing multiple-use management through a zoning approach, that is, a mosaic of resource uses that shift and rearrange across the landscape over time.

It is the dynamic, small-scale mosaic that makes this approach to multiple-use management so ecologically appealing. But these same features may also be what make this approach so politically impractical. Because the land allocations are dynamic, no decision is final. And depending on one's perspective or vested interest, there is always some alternative arrangement of land allocation that might work better than the one proposed. An atmosphere in which the Forest Service's motivations—and, at times, even its professional competency—are questioned makes it difficult for agency officials to render decisions and then make them stick.

A related alternative that has been discussed from time to time is a "dominant use" approach that designates larger areas to more limited uses with greater permanence. This option should not be confused with the "New Zealand solution" alluded to by Binkley in Chapter 5. In the 1980s, New Zealand dealt with its own forest management controversy by divvying up the landscape into one of two categories: exotic plantation forests, which were henceforth managed intensively for timber production by a public corporation established for this purpose, and natural forests, which were henceforth managed as preserves by a public agency. Even if this approach is proven to be ecologically sound over time, the existing forest land ownership pattern in the United States would require substantial transfers of public land into private ownership, and vice versa. Even very limited efforts of this kind, undertaken by public agencies as land exchanges to consolidate ownership and create more manageable landscapes, have rarely succeeded—and then only after lengthy negotiations, considerable expense, and public controversy. The philosophical concerns of some citizens over increased public land ownership are surpassed only by the concerns of others for the loss of multiple values on public lands when they are transferred to private ownership.

From a political standpoint, designating large areas—perhaps several national forests in a region—to be managed predominantly for timber, wilderness, recreation, or another use provides far greater clarity of purpose for local land managers. This alternative also narrows and clarifies the major beneficiaries, lending some of the positive aspects that Fairfax (Chapter 6) sees in the state trust lands approach. It also can provide greater continuity in management over time, particularly where the designation is done by law—as is done currently with federal wilderness areas—and thus would require new legislation to change the designated use.

However, while this approach could help bring about greater certainty and stability in federal land management, it severely restricts the flexibility of resource managers to adjust land uses over time in response to natural processes and events or to changes in social and economic factors. It also could result in a coarse and rigid pattern of forest environments that locally provide too much of one kind of habitat and too little of another, such as foraging habitat *and* nesting habitat for neotropical migratory birds, or summer range *and* winter range for elk and other large ungulates. Over time, this approach almost guarantees a diminished diversity of the existing animals and plants in the nation's forest ecosystems (Franklin 1993).

The triad model seems to offer a hybrid approach, with greater management flexibility than the dominant use model and greater decision-making certainty than multiple-use management as it has been practiced on the national forests. This model recognizes that there will be some areas of the landscape that are well-suited (by their productivity, natural resilience, and relatively low value for various nonmarket values) for intensive forest management to produce timber and other market outputs. Other areas may be well-suited (by their extraordinarily high biological diversity or other natural values) to serve as reserves, where there is a minimum of human intervention. (The extent to which intensified plantation management can be used to decrease development pressures on natural forests is estimated in Sedjo and Botkin 1997.) But the triad model still leaves the large majority of the landscape under active management as natural forest, or what Franklin calls the "landscape matrix" that holds the rest of the system together. Management of this natural forest would be characterized by a shifting mosaic of species compositions, successional stages, and physical structures.

Performance Measurement, Unrecognized Values, and Intuition

The problems Binkley (Chapter 5) identifies with natural forest management, particularly as it has been practiced on the national forests, are largely of values and the measurement of those values to ensure an efficient product mix from the forest. But Baskerville's claim (cited by Binkley) that "if you can't measure it, you can't manage it" seems to suggest that if a value isn't being measured, then it isn't being produced.

Like the planning decision models developed by the Forest Service, this philosophy implies a far more comprehensive understanding on these complex natural systems than we can rightfully claim. Like the inchworm measuring the marigolds, we often fail to see other values that are being produced, simply because our system has yet to devise a way of measuring or even recognizing these values. Less than one human generation ago—or less than one-quarter the age of the average mature forest—who among us understood the utilitarian values of biological diversity, or the critical role the world's forests play in protecting it? Only now are we discovering the role of forests in capturing and storing atmospheric carbon and the value this process represents in mitigating global climate change. What critical functions and services of forests will we discover in the future—values that we are still unable to recognize or appreciate, but that forests have been steadily and patiently providing for us for several million years?

Marion Clawson made tremendous and groundbreaking contributions to our efforts to value nonmarket benefits from forests, particularly outdoor recreation, and to incorporate these values in our optimization models. Nevertheless, it is important for us today to recognize that our knowledge is not perfect, and that the usefulness of such optimization models is limited by that imperfect knowledge. It is tempting to mistake the simulations from our models as truth, rather than as information provided by a decisionmaking tool that is at best only a rough approximation of the real world. As students and successors of Marion Clawson, we should attempt to derive valuations for biological diversity or carbon sequestration—as long as we keep in mind that, in real-world forests, we have yet to recognize and incorporate other important values in our models.

American society today is performance-measurement crazy. Yet, in an intuitive way, we seem to be able to implicitly acknowledge values that we cannot precisely see or measure. For a century now, despite all the controversies and conflicts among competing uses, Americans have continued to decide collectively, through our political system, that our national forests are worth what we pay for them. The much prolonged issue over what the national forests return to the federal treasury relative to annual appropriations is of far more interest inside the Capitol Beltway than elsewhere in the country. The millions of citizens who visit the national forests each year have their own systems for "benchmarking" and "performance measurement." Is the water still sparkling clear and cold? Are the rivers and lakes still full of trout, salmon, and other native species? Are the forests still vibrant and green, and do they still echo with birdcalls and the bugling of bull elk?

The extraordinarily competitive and entrepreneurial spirit that characterizes the American economy also characterizes our politics. It is inevitable in the free-for-all of American politics that a small minority of interests will continue to be disproportionately vocal in their efforts to sway the management of the national forests in one direction or the other. Still, the large majority of citizens seem so content with what the national forests are providing that they are unwilling to give up even a single acre. Certainly, any individual citizen can point out improvements that might be made, but Americans as a whole seem more satisfied with their national forests than the critics would suggest.

PINCHOT'S LEGACY

To paraphrase Mark Twain, rumors of the Forest Service's demise are greatly exaggerated. Despite voiced concerns that the agency no longer has a clear sense of mission or public purpose, people in rural communities across the country continue to regard the Forest Service as a good neighbor, a valued contributor to their economic vitality, and a reliable steward of their natural environment and quality of life. This constituency may be quiet, but it is deep, and it will continue to be the bedrock of support for the Forest Service and the national forests.

Another hundred years from now, if we're lucky, people like us still will be engaging in lively debates about the management of the national forests and the future of the Forest Service. In the long sweep of history, we may discover that Gifford Pinchot's greatest legacy to future generations was not the introduction of scientific forestry or even his articulation of the guiding philosophy of forest conservation and managed use. Pinchot's greatest legacy may be the 191 million acres of public forest land he caused to be conserved in perpetuity, to provide the greatest good, for the greatest number, for the longest time.

When we periodically come together to passionately debate what constitutes the just and proper use of these lands, perhaps we should pause a moment and be thankful that we have the national forests to argue over at all, and that they are of such value and beauty as to inspire this debate for the century now past and the century to come.

REFERENCES

Clawson, Marion. 1975. *Forests for Whom and for What?* Baltimore, MD: Johns Hopkins University Press for Resources for the Future.

Franklin, J. F. 1993. Preserving Biodiversity: Species, Ecosystems, or Landscapes? *Ecological Applications* 3(2): 202–5.

Hunter, M. L., and A. Calhoun. 1996. A Triad Approach to Land Use Allocation. In *Biodiversity in Managed Landscapes,* edited by R. C. Szaro and D. W. Johnson. New York: Oxford University Press.

Miller, Char. 1992. *Gifford Pinchot: The Evolution of an American Conservationist.* Washington, DC: Grey Towers Press.

Nelson, Robert H. 1995. *Public Lands and Private Rights: The Failure of Scientific Management.* Lanham, MD: Rowman and Littlefield.

NRC (National Research Council). 1990. *Forestry Research: A Mandate for Change.* Washington, DC: National Academy Press.

Pinchot, Gifford. 1998. *Breaking New Ground,* Commemorative Edition. Washington, DC: Island Press, 365.

Sedjo, R. A., and D. Botkin. 1997. Using Plantations to Spare Natural Forests. *Environment* 39(10): 14–20.

Seymour, R. S., and M. L. Hunter. 1992. *New Forestry in Eastern Spruce-Fir Forests: Principles and Applications to Maine.* Maine Agr. Exp. Sta. Misc. Pub. 710. Orono, ME: University of Maine.

———. Forthcoming. *Principles of Ecological Forestry. In Managing Biodiversity in Forest Ecosystems,* edited by M. L. Hunter. New York: Cambridge University Press.

Smith, Adam. 1796. *An Inquiry into the Nature and Causes of the Wealth of Nations.* Philadelphia: Thomas Dobson.

10

Changing Course

Conservation and Controversy in the National Forests of the Sierra Nevada

Lawrence Ruth

In the past forty years, environmental activism, public interest litigation, internal agency decisions, and legislative initiatives have changed the policies and management practices of the federal resource management agencies, in particular the U.S. Forest Service. Evolving scientific understanding of natural resources has intersected with broader social and legal developments. Initiatives for reform have been driven by two interrelated phenomena: political activism resulting from environmental and social concerns, and the incapacity of public institutions and private market forces to improve environmental conservation and management. As a result of these developments, management policy for the U.S. national forests—arising within the Forest Service and instigated by Congress through legislation—has been dramatically restructured.

Natural resource policy and planning initiatives in the Sierra Nevada region of California demonstrate the ways the national changes have played out in regional setting. The national forests of the Sierra Nevada[1] make a good case study for several reasons. First, a large proportion of the land area, approximately 42%, is national forest land. Second, the national forests of the Sierra Nevada contain most of the mid- to upper-elevation watersheds that supply water to urban California. In addition, these national forests offer a wide range of natural resources, furnishing

---

LAWRENCE RUTH is a visiting scholar in the Department of Environmental Science, Policy, and Management at the University of California, Berkeley. This chapter is an abridged, updated version of "Conservation on the Cusp," published in 1999 in *UCLA Journal of Law and Policy*, 18(1): 1–97.

timber for lumber and fuel, as well as minerals and forage for grazing. Finally, the region contains fish, wildlife, and varied opportunities for outdoor recreation. As demands for the many resources of the Sierra Nevada increased, so did conflicts among the timber industry, recreational users, environmental activists, and many others over the use of these national forests.

In this chapter, I explore public policies and issues associated with national forest management and examine their impact on the administration of the Sierra Nevada national forests from 1960 to 1999. I evaluate the progress of these initiatives and "rethink" the prospects for natural resource management and ecosystem conservation in the Sierra Nevada. First, I discuss the period from 1960 through the mid-1970s, when the Forest Service struggled to satisfy political and industry demands to increase use of forest resources, as well as to adapt to changing social conditions and to provide additional recreational opportunities; during this time, the agency responded to a series of national directives to develop management policies for the "multiple use" and "sustained yield" of forest lands and natural resources.

Next, I consider the environmental legislation and judicial decisions that reshaped administrative government and show how they unfolded in the national forests of the Sierra Nevada. I focus on the National Forest Management Act of 1976 (NFMA), which required the Forest Service to develop land and resource management plans for each national forest, and I examine the growing political activism and adversarial legalism that directly influenced the agency's activities in the Sierra Nevada.

Finally, I chronicle the continuing struggle by the Forest Service and others to find new methods to respond to conservation issues. In the 1990s, the Forest Service embraced the concept of ecosystem management and, gradually, recognized that the practice of ecosystem management required better knowledge of the landscapes, resources, and ecological dynamics. At the same time, communities and interest groups, frustrated by the lack of progress by the Forest Service, went outside the NFMA planning process and proposed innovations to improve planning and to speed the implementation of ecosystem management and conservation.

CONFLICTS, POLICY, AND LEGISLATIVE RESPONSE

The Roots of Conflict

For nearly eighty years after the creation of the forest reserves, as the national forests were originally known, land management policy was governed primarily by the Organic Act of 1897. This statute directed that the forest reserves be managed "to improve the forest within the boundaries, or for the purpose of securing favorable conditions for water flows, and to furnish a continuous supply of timber for the use and necessity of the citizens of the United States."

In addition, Gifford Pinchot, first Chief Forester of the Forest Service, had recognized that management and use of the national forests required a policy that would allow the Forest Service discretion to accomplish its goals while responding to local conditions (National Forest Service 1907; Wilson 1947).

During most of the Forest Service's first fifty years, the lack of detailed criteria for resolving conflicts over the use of national forest lands and resources was not a major concern. Where conflicts over uses or between users occurred, the agency often resolved them by public land designations or by adjustments to the specific project. After World War II, however, conflicts in the mission of the Forest Service became apparent, as did the agency's difficulties in responding to the demands of diverse users. A housing boom created an unprecedented demand for timber (Clary 1985, 94–112; see generally Dana and Fairfax 1980). When private timber supplies declined as a result of harvesting, the timber industry pressed for expanded sales of timber from the national forests (Clary 1985, 163). As the Forest Service increased timber sales to satisfy industry demand, conflicts between timber harvesting and other uses—particularly recreation—also increased.

Recognizing the increasing pressure to meet recreation needs, in 1958, Congress created the Outdoor Recreation Resources Review Commission (ORRRC) to review the situation and to make recommendations (see U.S. ORRRC 1962). The ORRRC submitted recommendations to Congress regarding increased public funding for recreational development and coordinated planning within agencies to provide better recreational opportunities. The Forest Service stood to benefit from the new emphasis on recreation, since implementing the ORRRC's recommendations would bring additional appropriations and permit the agency to better support the recreational opportunities that the public was demanding.

The Move Toward "Multiple Use"

As the number of visitors in national forests increased in the postwar period, the Forest Service sought legislation to confirm its authority to manage forest resources and forest-related activities so that it could meet the demand for recreation and wilderness preservation while continuing to provide for timber production and other commodity uses (U.S. Forest Service 1960; Dana and Fairfax 1980, 205).

In 1960, Congress enacted the Multiple Use Sustained Yield Act (MUSYA). The law stated that the national forests were to be managed for "the achievement and maintenance in perpetuity of a high-level annual output or regular annual output of the various renewable resources of the national forest without impairment of the productivity of the land." The statute also recognized the place and importance of various natural resource uses, including "outdoor recreation, range, timber, watershed, and wildlife and fish." This legislation enhanced the agency's authority to develop a variety of forest resources and activities appropriate to the needs of various users.

The Sierra Club was among those opposed to MUSYA, arguing that the Forest Service's strong commitment to timber production would undercut its support for other forest resources. Conservationists expressed concern that the Forest Service had become so focused on its role as a timber provider that it could not make fair decisions involving recreation uses or wilderness preservation (Dana and Fairfax 1980, 203–4).

Arguments by conservationists that timber considerations dominated the agency's agenda were common during the 1950s and 1960s. In this period, Congress considered the Forest Service timber management program to be consistent with its purpose, as outlined in the Organic Act of 1897, of furnishing "a continuous supply of timber for the use and necessity of the citizens of the United States." Encouraged by many powerful constituencies who wished to take advantage of the opportunities presented by national forests, congressional appropriations for the Forest Service overwhelmingly supported the agency's timber program (Clary 1985, 187; Hirt 1994, 234, 236–9). Accordingly, agency administration concentrated on building the timber program, and it had little reason to develop a management program more closely tied to a growing constituency whose interests centered on recreation or to search out policies that served a broader public interest.[2] As a result, the agency often appeared to have a bias toward timber production.

While the agency remained primarily attuned to timber interests, the professional foresters who made up the bulk of agency management and its professional staff were not always disposed to meet industry demands. By reason of their scientific and practical training as well as the conservation ethic of their profession, Forest Service officials possessed motivations and goals distinct from those of the timber industry. Despite the potential for tension between the two factions and occasional disputes over timber supply and harvest levels, working relationships between the Forest Service professionals and the timber industry became well established.

When Forest Service projects were opposed, often the potentially conflicting uses could be located separately. For example, timber could be harvested in one area while fishing, hiking, and other recreational activities could take place elsewhere. The diversity of opinions tended to cancel out each group's power, enabling the Forest Service to pursue its own agenda while claiming to have made compromises intended to satisfy each interest group (Culhane 1981; Foss 1960). The multiple-use rubric allowed the agency to avoid or defer many management controversies, but this approach could never fully satisfy important segments of the public.

Under MUSYA, the Forest Service experimented with planning to coordinate conflicting uses (Wilson 1978; Wilkinson and Anderson 1985). Each region prepared a regional multiple-use planning guide to steer local forest planning, and each national forest developed "Forest Land Use Plans" to guide the integration of various land and resource uses (U.S. Forest Service 1972; Hirt 1994, 222–3). Then, to tailor management principles to conditions, professional foresters prepared unit plans for watersheds or large drainage areas (U.S. Forest Service 1976; Leisz 1995). Unit plans translated national programs to the regions, which supervised the implementation of these policies on individual national forests; these plans tended to treat land uses and natural resources generally and did not conduct rigorous scientific investigations of the areas under consideration (Clary 1985, 172–3). This planning style demonstrated the utilitarian origins and professional norms of the "scientific movement" (Hays 1969, 2) of progressive-era conservation, which had become the guiding force in the professional development of U.S. foresters (Dana and Fairfax 1980, 52–3; Clary 1985, 6–17). The planning program was also a modern political exercise, ratifying the Forest Service's determinations of the "greatest good for the greatest number" (Wilson 1947). This system permitted timber sales to continue while allowing the agency to claim that it had become the nation's premier provider of outdoor recreation opportunities.

The Effort to Protect Wilderness Lands

From the early days of the Forest Service, the idea that large areas should be set aside for wilderness preservation had attracted some supporters. John Muir, and later Bob Marshall and Aldo Leopold (both of the Forest Service), as well as others, opposed the idea of resource use as the guiding principle for all forest lands (Hirt 1994, 35; see also Muir 1961; Dana and Fairfax 1980, 155–7). As the administration of the national forests matured, conservationists became concerned about the future of undeveloped and unspoiled areas in the national forests. By the 1930s, the Forest Service had already begun to restrict the use of certain areas and designate them as "primitive" areas (Dana and Fairfax 1980, 157–8).[3] However, wilderness advocates believed that if wilderness areas were accorded a status similar to any other use of forest resources, the remaining unspoiled areas would not be able to withstand pressure for access to timber and other commodities (Dana and Fairfax 1980, 157–8). In the postwar era, conservation groups wanted to ensure that existing primitive area designations would survive the timber industry's demands for increased timber harvesting. In 1956, they began to seek protection for wilderness through legislation that would make it impossible to administratively alter the status of these lands (McCloskey 1966).

In 1964, over the opposition of the Forest Service, Congress passed the Wilderness Act. Certain areas of the national forests, national parks, and other federal lands—including 2.1 million acres of land already protected by the Forest Service—were designated as "Wilderness" or "Pristine" areas. The agencies retained control over wilderness areas, but the new designation limited agency discretion in determining the disposition of these lands (Dana and Fairfax 1980, 227–9). The statute established a National Wilderness System, which did not permit timber harvesting on these lands but did allow for the continuation of other commodity-related activities, subject to presidential review. Under the legislation, other specified areas were to be studied and their status reviewed over the following decade. The new law increased protection for several areas in the national forests of the Sierra Nevada that were prized by users for their scenic beauty and recreational value.

For the wilderness advocates, the enactment of legislation after eight years of lobbying efforts represented a double victory. First, many areas were afforded more permanent protection. Second, the establishment of the wilderness system legitimized the philosophy of preservation-as-conservation as a component of future management decisionmaking for the national forests.

The National Environmental Policy Act

Enactment of the National Environmental Policy Act (NEPA) of 1969 represented a major watershed in public policy. Although the statute's provisions did not require that environmentally questionable projects be abandoned, it sought to ensure that decisionmakers and the public fully understood the environmental effects of proposed federal projects and programs. One major element was the requirement that an Environmental Impact Statement (EIS) be prepared for "major federal actions significantly affecting the quality of the human environment." Full disclosure of the environmental impacts of a proposed action ensured that the public had an opportunity to review the federal government's plans and proposed projects prior to their approval. The EIS also provided opportunities for individuals to comment on proposed actions, after which the lead agency could respond to comments and revise the project before reaching a final decision.

NEPA lacked substantive requirements for environmental protection; therefore, there was little initial expectation that preparing EISs would lead to dramatic changes in federal projects generally or in statutory programs—such as the multiple use and sustained yield policy—specifically. However, public disclosure of information through EISs provided citizens with an opportunity to mount formidable challenges to agency decisions in the political process and courts (Taylor 1984).

An early example of NEPA's effect and its importance to the environmental community was illustrated in the Mineral King controversy. The Walt Disney Company's proposed ski resort in Mineral King (a relatively undeveloped area located in what was then part of the Sequoia National Forest) was ultimately derailed by the insistence of environmentalists that the federal government comply with NEPA's procedures, even though the ski resort had been approved prior to NEPA. To satisfy opponents, an EIS was prepared for the proposed ski resort. The Sierra Club then challenged the adequacy of the document and sued to force consideration of the environmental effects on national park resources as a result of expanding the access road. Even though the lawsuit was eventually dropped, the delays created by the lengthy administrative and legal process ultimately caused the developer to lose interest in the project (*Ecology Law Quarterly* 1972; 1976). After several years, this area was transferred to Sequoia National Park.

Although the ecological significance of this result halted development in only one mountain valley and its environs, it was an important victory for conservation groups. The result greatly encouraged conservationists in the struggle against what they regarded as the tendency of the Forest Service to too quickly abandon its own conservation precepts in favor of a general compromise.

Timber Harvesting and the Monongahela Litigation

Faced with new demands on the national forests and growing conflicts between timber and other uses, the forest products industry, particularly in California and the Pacific Northwest, became concerned about Forest Service timber sale policies (Dana and Fairfax 1980, 199–202; Dowdle and Hanke 1985). As the demand for timber grew after World War II, cutting had increased on private lands in the western states. The availability of mature timber to harvest on these lands declined. Harvested lands were replanted for future use, but the new trees would not mature for many years. As a result, by the mid-1970s, many companies had become dependent on the national forests for timber supplies.

Timber interests were frustrated by the lack of a national strategy to respond to timber demand. The forest products industry pressed for improved planning and increased timber sales in the national forests to accommodate market demands (Clawson 1983; Dowdle and Hanke 1985). During the 1970s and 1980s, the Forest Service's Pacific Southwest Region harvested about 69% of the timber from the national forest lands available for harvest. In the industry's view, greater stability of the timber supply was essential. Many foresters maintained that harvest levels for the national forests, including those in California, had been set substantially below what the national forests could produce on a sustained yield level (Clary 1985, 165; Rey 1991; Zivnuska 1993). Foresters argued that an increase in harvesting would not jeopardize the long-term sustained yield of timber from the national forests (Clary 1985, 165).

In the Pacific Northwest, timber was harvested by clearcutting, which was associated with effects such as erosion and loss of forest habitat. However, in the Sierra Nevada, timber harvesting was accomplished mostly by single tree selection, whereby individual mature trees are designated for harvest and cut down (U.S. Forest Service 1976; Leisz 1995), and clearcutting was confined to a relatively small area. Eventually, foresters became concerned with the gradual decline in the health and vigor of trees that remained after repeated selection logging.[4] Although conservationists were aware of these effects, they preferred selection harvesting over clearcutting because it avoided the scarring and erosion associated with many clearcuts (Clary 1985, 180–185).

As timber harvesting and the number of visitors to the national forests increased, the effects of clearcutting and other silvicultural prescriptions became more evident, and more people expressed concern about the practice. Forest Service officials responded to complaints by touting the efficacy of clearcutting and minimizing the extent of the damage it caused. The agency argued that clearcutting was an appropriate part of a properly conducted silvicultural system, but economic efficiency was an important reason for support of the practice.

In 1973, a coalition of hunters, environmentalists, and others unhappy over plans to clearcut an area of the Monongahela National Forest in West Virginia brought suit to enjoin further clearcutting. The challenge centered on the interpretation of the Organic Act of 1897.[5] The Forest Service argued that the statute authorized clearcutting and several other timber harvest practices. In *West Virginia Div. of the Izaak Walton League of Am., Inc. v. Butz* [367 F. Supp. 422 (N.D. W.Va. Nov. 6 1973)], the court held to the contrary, stating that the Organic Act prohibited timber harvesting unless the trees were "dead, matured, or large growth" and individually "designated" and "marked" for harvest. Although the ruling did not prohibit clearcutting per se, it meant that clearcutting could not be used to harvest immature trees along with mature trees. Since the Forest Service was employing silvicultural management methods that were designed to do this, or did so by implication, the agency's system of timber management was effectively halted by the decision. This result was unexpected and stunned both the Forest Service and the timber industry. It was clearly unacceptable to the timber industry, which depended on the national forests as part of its available supply.

The Forest Service appealed the holding, but the Fourth Circuit Court of Appeals upheld the District Court's decision [*West Virginia Div. of the Izaak Walton League of Am., Inc. v. Butz,* 522 F. 2d 945 (4th Cir. 1975)]. In the wake of the Monongahela decision, environmentalists brought similar cases in district courts in South Carolina, Texas, Tennessee, Georgia, Alaska, and Oregon (Dana and Fairfax 1980, 317). The Forest Service faced the prospect of defeat in all of these cases and the loss of key management methods that it had come to rely upon. The holding marked a turning point in the agency's control over timber harvest methods in the national forests and marked the end to a period in which Congress afforded considerable deference to the Forest Service and its conception of multiple-use stewardship in the national forests. The future of the national forests and the role of the Forest Service were poised to become the subject of renewed public discussion and legislative debate. The discussion would occur in a highly charged and changing political landscape in which the environmental movement had acquired considerable ability to influence federal legislation.

Legislative Shift toward Strategic Planning

While the legal challenge to Forest Service authority worked its way through the court system, legislation was proposed to create a more reliable strategic planning framework for public and private renewable natural resources. The result was the Forest and Rangeland Renewable Resources Planning Act of 1974 (RPA). As a result of RPA, an assessment and program were prepared and forwarded to the legislative and executive branches of government. During the appropriations process, Congress reviewed the RPA reports, the assessment, and the program, but then its attention wandered.[6] As a result, the impact and significance of the RPA were negligible (Teeguarden 2000).

Nevertheless, RPA's strategic approach offered several valuable integrative mechanisms. The RPA related to three major Forest Service functions: administration of the National Forest System, forestry-related research, and agency responsibilities for providing technical and programmatic assistance to state and private forestry programs. Although these tools were never fully exploited, a strategic planning approach might have helped to realize shared goals for the conservation and management of public and private natural resources at national and regional levels. RPA provided broad strategic planning authority and is potentially relevant to emerging regional environmental planning and management initiatives (*San Francisco Chronicle* 1996)—particularly in regions such as the Sierra Nevada, where significant strategic planning and ecological issues of concern to both the state and federal governments extend across national forest boundaries.

THE NATIONAL FOREST MANAGEMENT ACT OF 1976

Enactment and Key Provisions

The Monongahela decision forced the Forest Service to modify its timber harvest practices to comply with the Organic Act. The agency quickly recognized that it lacked the statutory mandate to conduct a timber program on the scale that it had done previously and thus sought assistance from Congress. Legislation was crafted that entirely restructured national forest land and resource management planning. The agency's supporters in Congress, including Senator Hubert Humphrey of Minnesota, sought to restore Forest Service management authority and to reinvigorate the idea of multiple use by employing land and resource planning as the guiding principle for resolving natural resource management conflicts. During the legislative debate, Senator Humphrey rallied support for the statute by asserting that it would "get the practice of forestry out of the courts and back to the forests."

The National Forest Management Act (NFMA) of 1976 sought to increase Forest Service responsiveness to environmental values and required that the agency use economic analysis as part of its decision-making criteria. The statute that emerged was an attempt at compromise, calling for the implementation of natural resource planning that would reconcile the public demands related to conservation with the need for timber production and other commodity interests. NFMA directed the agency to develop plans that would respond to the resource conditions encountered in each national forest, and it allowed a range of forest management activities and land uses substantially the same as those prior to the Monongahela decision. Under NFMA, "multiple use and sustained yield" of forest resources remained the focus of national forest management while the public was assured that conservation objectives were to be treated seriously.

NFMA implicitly promoted planning as a means to better ensure a stable management environment. This approach pushed the Forest Service to develop new forms of professional competence and interdisciplinary expertise to address environmental issues. The statute allowed clearcutting only where it could be shown to be the "optimum" silvicultural method.[7] It also mandated the conservation of biological diversity, requiring the Forest Service to ensure that its plans would provide for "minimum viable populations" of forest species. The act also established procedures requiring the coordination of forest planning, environmental assessment, and public comment on agency planning proposals.

NFMA fundamentally restructured public land planning to produce more balanced plans and to reduce the likelihood of legal battles by incorporating two elements central to the broader administrative reform of the era (Stewart 1975). First, the relationship between law and administrative behavior is specified in the statutory elaboration of the planning process (Wilkinson and Anderson 1985). Second, NFMA expanded opportunities for public involvement in the planning process, seeking to permit unprecedented levels of public participation in management decisions. The close relationship between law and administrative behavior was intended to ensure that planning decisions were consistent with the law and that the agency would make its reasons for management decisions explicit in the plans themselves. National forest resource planning was to be undertaken pursuant to detailed statutory instructions to ensure that adequate consideration was given to both resource protection and development.

NFMA planning and the NEPA process prompted the Forest Service to consider information that previously was undervalued or ignored. For instance, NFMA directed the agency to use an "interdisciplinary team," which consisted of agency scientists and resource professionals with diverse scientific and professional skills. By requiring input from new kinds of experts, NFMA intended to make certain that the condition and sustainability of forest resources were given full consideration during agency decisionmaking. The new procedures meant that detailed analysis and documentation would be required prior to an agency action. This difficulty was compounded in the case of the national forest lands because NFMA called for standardized—but also site-specific—analysis and planning for a set of extremely varied lands and natural resources (Dana and Fairfax 1980, 328–36).

In addition, the law vested Forest Service administrators with authority to make planning decisions within a range of legally acceptable outcomes that would achieve the greatest "net public benefit." Inherently, decisionmakers were left with a great deal of discretionary authority. Land management planning, however, like many other public programs, was conducted in a highly charged political environment. Therefore, NFMA also sought to increase direct public input into agency decisions. Public involvement was intended to reorient administrative decisionmaking from a strict reliance on expert management and toward a process resembling a political dialogue between the administrator and the public (Reich 1985; Friedmann 1987; Handler 1988; Wondolleck 1988). From the early days of land management planning, a succession of executive branch appointees paid close attention to the possible implications of the agency's decisions. The Forest Service was therefore expected to act with both technical proficiency and sensitivity to public and political opinion.

NFMA planning was introduced into an already contentious atmosphere, with timber interests and environmentalists fundamentally opposed to each other's position. New controversies occurred over natural resource issues and areas where a strong tide of activism had already left an indelible mark. In light of these complexities, the expectation that NFMA planning would retain sufficient flexibility for managers to respond to varying local needs and conditions was perhaps a forlorn hope. NFMA's land management planning mission was subject to continuing scrutiny by activists, interest groups, and scholars. At the time of the new law's enactment, many analysts were skeptical about its power to overcome the polarization between industry and environmentalists. Many familiar with contemporary public land management in the United States had come to view controversy as the normal condition for public land policymaking. To many observers, the effort to blend conflicting aims, simultaneously promoting multiple use and sustainability, was a prescription for increased conflict (Stroup and Baden 1983; Rosenbaum 1984; O'Toole 1988). As the planning process got under way, policy scholars expressed added doubts about the chances for successful culmination of planning, especially in light of NFMA's procedural elements that offered opponents many opportunities to challenge the implementation of agency plans (Behan 1981).

NFMA Planning in the Sierra Nevada: Initial Effects

In the late 1980s and early 1990s, to implement NFMA planning requirements, the Forest Service produced Land and Resource Management Plans (LMPs) for future resource uses within each national forest in the Sierra Nevada.[8] Planning focused on individual national forests, with little regard for regional factors or characteristics. Nevertheless, NFMA planning represented a major undertaking for the agency, and several elements of the plans illustrated significant departures from the policies that had previously guided national forest management in the Sierra Nevada. Armed with statutory language that once again permitted the agency to utilize even-aged management and clearcutting, the plans called for increased timber harvesting using these and related silvicultural methods[9] in many regions, including the Sierra Nevada, despite the controversy. The LMPs also proposed significantly increased timber harvests in the Sierra Nevada and the continuation of selection logging, though at somewhat reduced levels. The rationale for the change in harvesting methods was to ensure an even distribution of tree age classes across each forest. Planning documents presented to the public suggested that as these stands were cleared and replanted, overall growth would increase, allowing the forest to continue to supply relatively high amounts of timber (U.S. Forest Service 1988).

Individual forest plans also considered the status and eventual use of remaining roadless areas. Forest Service planners viewed those areas with the capacity to produce timber as potential sites for intensive timber management (U.S. Forest Service 1990c). This strategy fit well into the agency effort to improve timber yield. Similarly, in other areas, the natural mix of species had been or was being eclipsed by the growth of white fir as a consequence of earlier timber harvests and fire suppression. These stands were to be harvested and replanted with species that existed before human intervention. This strategy was one method of restoring the vitality of the forest but also was designed to increase commercial timber yields on the remaining commercial timber base. Critics, however, questioned the entire premise behind the plans—that timber production should take precedence over other forest uses and values. They viewed the Forest Service's description of the benefits of these silvicultural methods as insufficient to justify increased harvesting or clearcutting.

The new policy produced some dramatic changes. For example, timber was harvested by clearcutting immediately adjacent to a number of giant sequoia groves located in Sequoia National Forest. These clearcuts were among the most visible changes implemented as a result of land management planning. They were justified in part by Forest Service managers as a method for enhancing the vigor and promoting regeneration of giant sequoias (Solnit 1997). Environmentalists, however, viewed the timber sales and harvests as evidence of irresponsible stewardship in these relatively rare ecosystems (Stewart and others 1992; see also U.S. Forest Service 1994). They regarded the logging initiative as proof that the agency was more interested in obtaining timber than in protecting the landscape and sued to stop additional timber harvests near the groves (Fisk and others 1997). Clearcutting in these areas eventually ended, and the management of the groves was restructured as part of a negotiated settlement of the dispute (U.S. Forest Service 1990a).

Responses to Plans for the Sierra Nevada

As the Forest Service completed LMPs for most of the national forests of the Sierra Nevada, the plans were carefully scrutinized, and the proposed increases in the intensity of management practices quickly attracted critical comments from many segments of the public, including environmentalists, timber interests, scientists, area residents, and others familiar with the region.

A core of environmental activists emerged during the planning process. They opposed the use of clearcutting; sought protection of the forest environment, particularly in the remaining roadless areas; and lobbied for recreation and other noncommodity forest uses. Many found it ironic that, following upon earlier struggles that culminated in the Monongahela decision, they should be forced to fight against clearcutting again (Beckwitt and others 1986). As the activists grew more sophisticated and better organized, they developed both the facility and predisposition to question agency proposals.[10] Many activists associated themselves with environmental organizations that had experience in national forest management issues, such as the Sierra Club, the Wilderness Society, the Natural Resources Defense Council, the Audubon Society, and the National Wildlife Federation. The national organizations, for their part, depended to varying degrees on local groups and representatives for information about individual forests and for alerts about the ramifications of the plans (U.S. Forest Service 1990b).[11]

Timber interests also paid close attention to the planning process in the Sierra Nevada. The timber industry, which was sympathetic to state and local officials, and others requested that the Forest Service adopt a regional policy that set timber harvest levels closer to the RPA strategic targets for commodities. They also requested that the Forest Service adopt alternatives in the final plans for the national forests that more closely conformed to RPA targets (Craine 1986). The Forest Service considered these comments and referred to them in responses to public comments on planning documents. Many Forest Service officials were sympathetic to the tenor of the industry's comments but had come to realize that RPA's timber harvest targets for the Sierra were probably unachievable because of a combination of environmental constraints and public preferences for other forest uses.

The State of California also responded to the plans, commenting on both the process as a whole and individual plans. Comments by Department of Forestry and Fire Protection went beyond discussing timber harvest levels and plans for individual national forests (Forest and Rangeland Assessment Program 1988). From the state's perspective, these issues required full consideration of the panoply of demographic, social, and environmental issues that affected the national forests and the surrounding landscape. These comments were intended to reorient Forest Service planning toward a more integrated consideration of the national forests and their contribution to the region and to the state. The department's response sought to better coordinate land management planning with what was occurring on adjacent private lands. For example, the state suggested that sustained timber yield be calculated by using a regional timber inventory, assessing the timber stocks on both public and private lands to develop a regional sustained yield level, rather than a level for a single national forest (Ewing 1992).

State concerns extended to a variety of noncommodity issues, seeking to draw the Forest Service more deeply into planning for watersheds and regions that consisted of multiple national forests. Despite administrative regulations stating, "The responsible line officer shall coordinate regional and forest planning with the equivalent and related planning effort of other Federal agencies, State and local governments, and Indian tribes," the Forest Service made little or no effort to comply with the state's suggestions. The agency did not take the state's comments as a serious criticism of land management planning, explaining that its statutory mandate did not grant it authority to undertake the mission that the state proposed.

As the scope and intensity of the agency's proposals were laid out in the plans, it became clear that the natural character of many areas in the national forests would change drastically. In response, local, regional, and national groups organized and worked together to oppose elements of the plans to which they objected. Predictably, some of the same individuals and organizations that the Forest Service faced in earlier struggles over clearcutting and wilderness issues resurfaced. In its defense, the agency emphasized its compliance with the procedural rules that had been established for planning. The Forest Service maintained that the value of its vision of comprehensive land and resource planning would become evident over time, forcing observers and opponents to acquiesce to the results of planning. Although the Forest Service believed that its land management plans were legally acceptable, the now-familiar aspect of environmental opposition to agency proposals was a troubling sign that the Forest Service might have difficulty getting forestry entirely "back to the forests." This result became more evident as land management plans were completed and adopted as the guiding policy for management of the national forests of the Sierra Nevada.

Conservation of the California Spotted Owl

Legal requirements for protecting plant and animal species have played an important role in the evolution of federal and state natural resource management. Scientific analysis has helped administrators make decisions consistent with the law but often has revealed the shortcomings of established practices. Scientific information in combination with the mandates for the protection of sensitive species essentially required the Forest Service to take account of the needs of the species, to provide for their habitat requirements, or to modify any land management plan that fails to do so. Accordingly, the information gathered in the NFMA planning process in the Sierra Nevada meant that management practices, some in use for many years, might no longer be acceptable. The case of the California spotted owl highlights this challenge.

The range of the California spotted owl (*Strix occidentalis occidentalis*) extends through portions of the Sierra Nevada, within national forests in the Pacific Southwest Region (Region 5) of the Forest Service. A related subspecies (the northern spotted owl) inhabits forests in Oregon, Washington, and northern coastal California—within the Pacific Northwest and the Pacific Southwest Regions of the Forest Service (Verner and others 1992, 55). In the 1980s, the Forest Service initiated management programs for protecting small areas of northern and California spotted owl habitats in a gridlike pattern known as Spotted Owl Habitat Areas (SOHAs), starting in seven national forests in the western Sierra Nevada in 1981. The SOHA strategy permitted limited timber harvesting in parts of SOHAs not immediately adjacent to nest trees. Lands outside of SOHAs also were used for nesting, roosting, and foraging by the owls, but the SOHA policy did not affect timber harvests on the remainder of forest lands (U.S. Forest Service 1993b).

Research conducted on the northern spotted owl raised the possibility that the SOHA strategy did not sufficiently protect the owl's habitat and that the continued use of clearcutting was detrimental. The Interagency Scientific Committee reviewed the condition of the northern spotted owl in 1989 and concluded that the SOHA management strategy would not sufficiently ensure the survival of the species and that continuing the strategy would lead to further decline in the owl population (Thomas and others 1990). The Fish and Wildlife Service listed the northern spotted owl as a "threatened" species under the federal Endangered Species Act of 1973 (ESA) in June 1990. Research findings suggested that the SOHA policy and subsequent administrative actions used to protect the habitat of the California subspecies were also likely to be found inadequate.

Few demographic or ecological studies were specific to the California subspecies. The lack of biological information made it difficult to offer guidance as to what type of habitat management should be adopted and to justify any significant change in management guidelines. Nevertheless, scientists, managers, and the public were concerned that extensive clearcutting might jeopardize the species' survival.

Eventually, the Natural Resources Defense Council appealed timber sales in areas of the Sierra Nevada used by the spotted owls adjacent to the SOHAs (NRDC 1991). The administrative appeal argued that

NFMA's mandates—especially the section of the regulations requiring the Forest Service to ensure that its plans would maintain "viable populations of native and desired nonnative vertebrate species"—required the agency to take greater steps toward protecting wildlife habitat. The Forest Service determined that the argument had substantial merit and decided to try to resolve the issue by changing its policies without waiting for the results of the administrative and legal processes.

Developing a Successful Conservation Strategy

In May 1991, state and federal agencies convened the California Spotted Owl Assessment and Planning Team to assess the status of the owl and to explore alternative strategies to conserve the subspecies and its habitat. Forest Service and other scientists were appointed to a "Technical Team" and asked to analyze the status of the owl. The team investigated the loss of suitable habitat in the Sierra Nevada, evaluated several possible alternative management strategies for the owl, and presented its analysis and recommendations in May 1992. The team concluded that habitat loss had been caused by even-aged silvicultural practices and catastrophic fire (Verner and others 1992). The analysis suggested that existing policy and management measures used to protect the spotted owl and its habitat were completely inadequate. The team also noted that the recent LMPs called for increased clearcutting[12] and other forms of regeneration harvests that removed large-diameter trees, which were the preferred nesting and brooding sites of the owl.

The Technical Team clearly regarded current management direction as detrimental to the long-term well-being of the habitat and the species (Verner 1993). Under the LMPs for the Sierra Nevada national forests, the team estimated that the amount of suitable habitat would further decline at a rate of 229,000 acres per decade (Verner and others 1992). The research also concluded that suppression of fire had accelerated the accumulation of fuels and significantly increased the likelihood of fires that would destroy timber stands, including those essential to the spotted owl. The team concluded that owl habitat in the Sierra Nevada could not be protected while allowing clearcutting or otherwise permitting the removal of large old-growth trees from these forests.

To remedy the deficiencies of existing management policy, the Technical Team proposed an interim strategy for protection of the California spotted owl. Known as the "CASPO" (California spotted owl) recommendations, the strategy called for additional research but also recommended an end to the harvesting of large trees in areas used by spotted owls in the national forests of the Sierra Nevada. The Technical Team's strategy also recommended that the Forest Service begin to utilize more aggressively a variety of fuel management techniques, including the thinning of dense forest stands (Verner and others 1992).

To incorporate the recommendations into management policy, the Forest Service prepared the *California Spotted Owl Sierran Province Interim Guidelines Environmental Assessment* (U.S. Forest Service 1993a). The environmental assessment, and the subsequent decision by the regional forester in January 1993, incorporated substantially all of the Technical Team's management recommendations into an interim strategy for managing the national forests of the Sierra Nevada (U.S. Forest Service 1993c). Thus, the legal requirement that the agency provide for "minimum viable populations" of forest species significantly altered the Forest Service's course. Earlier changes adopted during NFMA land and resource management planning (such as the use of clearcutting in many areas of the Sierra Nevada and harvesting of large-diameter timber in forest areas favored by the owl for nesting and foraging) were suspended. An independent estimate predicted that implementing the recommendations of the Technical Team would reduce timber harvest in the national forests of the Sierra Nevada dramatically (Ruth and Standiford 1994, 10, table 4-4). When the recommendations were implemented, harvests were reduced by approximately one-third (U.S. Forest Service 1998b; Stewart 1996).

Summary of NFMA's Effects

NFMA planning, conceived as a method to provide for multiple use of the forests while ensuring resource sustainability and conservation of biological diversity, clearly did not end conflict over the future of the national forests. The law's provisions channeled the political activism of the era and objections to the plans into the legal process. Conflicts over forest use and protection increasingly came to be characterized as scientific questions that, under the law, could provide a basis for a legal challenge to the plan.

As the Forest Service's land and resource plans were completed in the Sierra Nevada, the Forest Service did not evolve methods to respond to the substance of public objections to NFMA planning. At first, the agency's answers to its opponents in forest planning disputes were formal and contributed to the climate of adversarial legalism that surrounded forest planning. Public pressure on the Forest Service to reinterpret its statutory responsibilities, along with the likelihood of legal challenges, forced the agency to reexamine the emphasis it formerly had placed on commodities. Only at this juncture did the agency begin to seriously restructure natural resource planning and management to better incorporate scientific information about local areas into national forest planning and management (see, for example, Yaffee 1994).

ECOSYSTEM MANAGEMENT IN THE SIERRA NEVADA

The Forest Service, despite its policy modifications, remained under siege. Challenges to its competence, authority, and mission forced the agency to explore new methods to respond to resource conservation issues. The agency's treatment of the Sierra Nevada was influenced by its experience regarding the spotted owl in the Pacific Northwest, where the dispute had been played out in a series of drawn-out legal battles. Stung by those legal defeats, the agency rapidly began to change existing policies in California. The Forest Service embraced the concept of ecosystem management to respond to the challenges posed by environmental and ecological issues (Robertson 1992) and embarked on policy initiatives designed to avoid a repeat of defeats suffered elsewhere, believing that this approach would satisfy the panoply of legal directives that required the agency to take account of increasingly complex ecological information.

The Sierra Nevada Ecosystem Project

The Forest Service and its critics recognized that the practice of ecosystem management required better knowledge of landscapes, resources, and ecological dynamics. Congress began to pay attention to resource managers, environmentalists, and scientists who called for action to resolve controversies over old-growth forests. Grassroots environmental interests, later joined by the Forest Service, called for the development of an independent map of old-growth forests in the Sierra Nevada.[13] Some agency personnel suggested that what really was needed was a thorough review of all resource issues associated with these forest areas and ecosystems. Congress eventually appropriated a modest amount to the Sierra Nevada Ecosystem Project (SNEP), an independent panel of scientists who would conduct a scientific study of the remaining old growth in the national forests of the Sierra Nevada in California and further analyze the entire Sierra Nevada ecosystem.

The legislative background of the study emphasized that the report was to advise Congress rather than to prepare a plan or to develop a spectrum of alternatives for future consideration in an EIS. The study's objective was to assess scientifically the health and sustainability of the Sierra Nevada. Pursuant to Congress's request, the project was to examine existing research to provide an overview of the status of resources and ecosystems. The final report was also to include strategies to protect the health and sustainability of the Sierra Nevada in the future while providing resources to meet human needs. The general conception of the study was that it would be informed by Forest Service research data and expertise but insulated from agency influence. The agency, however, had already committed itself to a revision of existing plans to better conserve owl habitats, and the results of an independent study would furnish needed information about the condition of additional species and other natural resources.

SNEP built on existing scientific research, conducted assessments of the region's ecosystems and natural resources, and examined society's relationship to them. The mission and the objectives of the SNEP team were separate and distinct from the traditional approach to land management planning under NFMA. SNEP was conceived as a bioregional assessment, not a management or a decisionmaking process. Rather than offering management alternatives, SNEP offered "strategies" that would illustrate alternative possibilities for specific ecosystem attributes. The strategies were intended to highlight trends affecting the future of the region and its ecosystems and to educate Congress and the public. The *Final Report of the Sierra Nevada Ecosystem Project* was published in 1996, with an addendum in 1997 (Centers for Water and Wildland Resources 1996).[14]

The Progress of Forest Service Management Proposals in the 1990s

As the work of the SNEP progressed, Forest Service management plans for the Sierra Nevada again attracted attention. Timber interests and several counties whose economies and revenues were affected by the regional forester's 1993 decision implementing the CASPO strategy were roundly displeased with the new policy. They challenged the decision in court, but the "interim strategy" proved sound enough to withstand these challenges.

The CASPO report had recommended that measures be taken to reduce accumulations of forest fuels in and around forest stands in order to reduce fire danger in the national forests. Without timber harvesting, the Forest Service had fewer opportunities to do so. Over the previous several decades, fuels accumulations in many areas had been reduced as the area was logged—although logging did not eliminate fire risk or prevent fires or from spreading.[15] On the other hand, statutory arrangements had designated a portion of timber harvest receipts for use in the forest where the sale originated. Fuels management projects, especially in areas that required independent or additional treatments, depended in part on these funding sources. As logging levels dropped, projects that once might have been possible became problematic or no longer viable.

Finding and implementing solutions to the practical problems of forest and fuels management became increasingly difficult for the Forest Service. Prescribed burning, which can reduce the fire danger by controlling fuels accumulations, was used on a very limited basis in the Sierra

Nevada region. Due to the effects on air quality, risks to human safety and to property, and considerations of potential liability, these programs require careful planning and supervision, and they are costly to administer. They also require extensive coordination between local government, resource managers, and air quality management agencies (*San Francisco Chronicle and Examiner* 1999). For these reasons, prescribed burning is difficult to implement on a scale large enough to make it effective for managing fuels in the national forests of the Sierra Nevada (Centers for Water and Wildland Resources 1996, vol. 1, chap. 3).

In January 1995, after review of the results of ongoing research on the spotted owl, the Forest Service issued a new plan for conservation and management of the national forests of the Sierra Nevada. The plan and the accompanying *Draft EIS: Managing California Spotted Owl Habitat in the Sierra Nevada National Forests of California* (DEIS) (U.S. Forest Service 1995) were intended to help the Forest Service implement a long-term policy for forest management in the Sierra Nevada. Scientific reviews and public comment on the DEIS were swift and extremely critical. They claimed that the document had not taken proper account of research findings pertaining to old-growth forests and other ecosystem resources and did not meet public expectations for an ecologically sensitive plan (Franklin and Fites-Kauffman 1996, vol. 2, chap. 21; Franklin and others 1996, vol. 2, chap. 3). The agency substantially revised the document in response to these concerns and circulated it for additional public comment. The later draft, the *Revised Draft Environmental Impact Statement for the Sierra Nevada National Forests* (U.S. Forest Service 1996), was released in August 1996. Even before its release, this draft was attacked for many of the same deficiencies and met an even swifter demise.

To get the agency on track, the Clinton administration requested a formal review of recent scientific research and of the adequacy of the revised draft EIS prepared by the Forest Service. The California Spotted Owl Federal Advisory Committee conducted an extensive review of Forest Service planning and decisionmaking. The advisory committee found that agency planning had become too narrowly focused on spotted owls to the detriment of other species and forest attributes; that the agency had not adequately considered available scientific information pertinent to spotted owls in formulating its management plans; and that the agency had failed to account adequately for other resource concerns, particularly fire and fuels management, old-growth forests, and the needs of other sensitive species (USDA 1997).[16]

In 1998, the Forest Service initiated the Sierra Nevada Conservation Framework to further revise management policy (U.S. Forest Service 1998a). A draft EIS—technically, an amendment to the land and resource plans of each national forest in the Sierra Nevada—was released for review and comment in May 2000.[17] The new plan, incorporating research from SNEP, the California Federal Advisory Committee, and other studies, is intended to chart a new course for environmental stewardship in these national forests. The plan's modifications will have less of an effect on the timber industry than the changes arising as a result of the regional forester's 1993 decision to implement the CASPO strategy and to significantly reduce timber harvest levels accordingly. Nevertheless, the Forest Service must address other controversies in the Sierra Nevada Conservation Framework EIS. These issues include fire and fuels management, conservation of old forest ecosystems, the needs of wildlife species, and conservation of aquatic, riparian, and meadow ecosystems. Accordingly, further changes in land management practices are anticipated after the completion of the final EIS in fall or winter 2000.

Simultaneously, the status of the California spotted owl remains under investigation. Research completed after the report of the Technical Team was released indicates that the spotted owl population in the Sierra Nevada continues to decline. The Sierra Nevada Forest Protection Campaign and the Southwest Center for Biological Diversity have petitioned for a review (Knudson 1999; Southwest Center 1999; Martin 2000). If a review is conducted, then the California species could one day be listed pursuant to the ESA as "threatened" or "endangered," an action that could require still further modification to planning initiatives for the Sierra Nevada national forests.[18]

Policy Prospects for Ecosystem Management

Until recently, the Forest Service served different segments of the public by providing for their needs separately (Hirt 1994, 222–3). With increased demands from and use of the forests by the public, it became more difficult to successfully serve one constituency without angering another (Centers for Water and Wildland Resources 1996, vol. 1, 51–2). Although NFMA was designed to address this dilemma by blending resource use and conservation to respond to the natural attributes and public values within each national forest (Wilkinson and Anderson 1985), the Forest Service had only limited success in incorporating an ecological approach into land and resource planning and management. The agency's ecosystem management initiative, which attempted to integrate the diverse statutory mandates governing its land and resource management activities (Robertson 1992), represented a somewhat tardy response to citizen concerns about the ecological effects and implications of Forest Service resource management activities. Concerns expressed by agency personnel often had been overlooked or ignored in earlier management plans. Environmentalists and scientists cautiously welcomed the changes, but agency operations continued to be perceived as inconsistent with scientific understanding and the legislative mandates for the management of the national forests (NRDC 1993). For these reasons, the implementation of ecosystem management in the Sierra Nevada thus far has proven unsuccessful in defusing conflicts.

Despite the promise of ecosystem management, public dissatisfaction and political and social activism over national forest management have continued. Citizen activists and interest groups draw on specific elements of NFMA and its regulations, such as the requirement that planning protect "minimum viable populations" of forest species, that lend support to their particular positions (NRDC 1993). The agency, on the other hand, must try to balance the operation of particular provisions with other goals of NFMA. In many instances, it has done so only to find that a decision will not meet legal or regulatory standards. Conflicting objectives, such as habitat conservation and commodity production, are supported by both elements within NFMA and other applicable laws (Sedjo 1999). To achieve often divergent goals, the Forest Service promulgates policy intended to reconcile the conflicts within the framework of planning, conservation, and management. Absent strong criteria to defend its choices, attempts to balance competing objectives are vulnerable to legal challenges because they do not satisfy specific provisions of the law or regulations. If the agency's decisions reflect inattention to legal mandates pertaining to environmental protection, then public and scientific scrutiny quickly translates into opposition, administrative appeals, and lawsuits. The larger outcome of the NFMA processes—unintended by the Forest Service, but consistent with judicial interpretation of NFMA and NEPA—is that plans and management activities poised to have a substantial and often adverse impact on the ecosystem have been prevented from being fully implemented (Centers for Water and Wildland Resources 1996).

Citizen activism helped the forest planning process function as a self-correcting mechanism, albeit an awkward one, to remedy administrative errors (Yaffee 1994). It has had lasting and positive effect. Although its effect on the implementation of management plans and the production of timber and other outputs may appear to be detrimental, activism has been beneficial in upholding and enforcing the provisions of the law.

The Forest Service's ongoing effort to change its policy to incorporate new scientific information about the California spotted owl is one example of how activism has worked in practice. After Forest Service plans for high-level timber harvesting in the northern spotted owl's habitat were blocked repeatedly in court, the agency made major policy changes to protect the owl species in the Sierra Nevada. The 1993 shift in Forest Service policy regarding the California spotted owl changed the focus of planning and management from commodities to conservation, focusing first on habitat and then on ecosystem resources. The environmental impact reporting requirements of NEPA were essentially given a second chance to operate. As the agency fought to avoid listing the owl as a threatened or endangered species, the focus of planning and management became more detailed and more dependent on site-specific information. After completion of CASPO, SNEP, and subsequent studies, the new goals of Forest Service planning became protection of the spotted owl, conservation of biological diversity, and providing for resource sustainability in the Sierra Nevada.

The Quincy Library Group

The Quincy Library Group (QLG), based in the northern Sierra Nevada, was originally formed in 1993. In Quincy, even before the CASPO decision was announced, individuals who represented different sides of the debate began discussions to determine whether they could agree on any aspects of resource management and forest conservation (Bernard and Young 1997). Bill Coates, then a Plumas County Supervisor; Michael Jackson, a local attorney and environmentalist; and Tom Nelson, a professional forester employed by Sierra Pacific Industries (the largest timber company in the state and owner of a local lumber mill), participated in the first meetings, which were held in the Quincy Public Library. As the discussions continued, the group expanded and gradually came to include local citizens, members of the local environmental community, and representatives of timber interests—all of whom represented a variety of experience and perspectives.

QLG discussed methods for managing and restoring forest ecosystems while providing a means to sustain the community and the economy. The objective was to develop and lobby for the implementation of a management plan that incorporated scientific information and otherwise complied with environmental laws such as NEPA, ESA, and NFMA. QLG explored resource management alternatives for the Plumas National Forest, the Lassen National Forest, and the Sierraville Ranger District of the Tahoe National Forest.[19]

Understanding the swift and significant changes in timber harvesting policies caused by the CASPO decision in 1993 is the key to understanding the objectives and results of QLG. The effect of CASPO was to eliminate clearcutting and the harvesting of old growth and other large trees in adjacent national forests for the foreseeable future. The significantly lower timber harvest levels caused rancorous debates about the effect of the policy change and related issues (Duane 1998, 791). Timber interests and supporters emphasized the efficacy of clearcutting and sought to justify the harvest of larger old-growth trees as a vehicle to continue brush control and otherwise improve fuels management. On the practical side, however, the CASPO decision temporarily ended the legal and administrative controversies over these issues in the national forests of the Sierra Nevada.

The new policy decision had ancillary economic effects. Under an old formula, federal law mandated that the Forest Service distribute 25% of its gross receipts to the counties of origin.[20] In Quincy, the CASPO initiative was expected to reduce local government revenues, school financing, local employment, and business activity (Jackson and Nelson 1996). The policy change created or highlighted many social, economic, and resource management issues, such as social well-being of the local community, employment for area residents, and financing for local government and schools (Jackson and Nelson 1996; Doak and Kusel 1996; Kusel 1996a). In light of the resource management constraints imposed by CASPO-related policies, traditional methods of financing much of the practical work of forest management were no longer available.

With the decline in timber harvesting, the Forest Service reduced its staff. The agency also lost much of the funding it previously had obtained directly from timber sale receipts. These developments limited many options for on-the-ground activities, including fuels management. In addition, pending the completion of further research on the spotted owl, the pace of the agency's management activities had slowed dramatically. As a result, it had become difficult to undertake fuels management projects, because a large proportion of this work had in the past been addressed either directly or indirectly as a function of the Forest Service's timber harvesting program (Ruth and Standiford 1994, 25–27).

Several aspects of the QLG's initiative differed dramatically from traditional modes of Forest Service planning, conservation, and management. First, the diverse experiences of the group's members in the forest planning process and in resource and environmental issues in the region—as well as the members' knowledge of each other's effectiveness as adversaries—brought experience and pragmatism to the discussion table (Terhune 1999, 4).

Second, QLG built a local coalition entirely outside of the ordinary channels of public involvement in the NFMA planning process. QLG was and is guided by the direct involvement of the individuals represented at the group's inception. As a result, some national and local interests, particularly environmental groups, believe their voices have been excluded from the opportunity to influence the group's decisions (Edelson 1997), despite the presence of several environmentalists on QLG's steering committee. Initial meetings actively excluded Forest Service personnel, although this changed over time. Local grazing interests also expressed irritation that QLG has been slow to respond to their concerns (Terhune 1999, 10), whereas Sierra Pacific Industries, the state's largest timber company, has been an important player in the group. In contrast, efforts by other local partnerships excluded the players with key economic interests in the outcome of planning from their groups and away from the planning process (Nijhuis 1999). For QLG, however, industry involvement was an essential if implicit ingredient, inasmuch as the plans developed by QLG were meant to keep the local mill in operation.

Third, QLG was able to look at the practical aspects of resource management and the feasibility of resource management methods and tools. For example, after a forest fire burned more than 44,000 acres in the area near Quincy in the summer of 1994, the community became particularly concerned about fire danger (Marston 1997). Drawing on SNEP and other sources, QLG sought to design fuels management projects that would gradually return fire to a more natural role in the forest ecosystem. However, after examining the entire process of fuels management, the group determined that the relatively high costs for fuels management projects would constrain fuels treatment in the area.

The group's pragmatism expressed itself in an entrepreneurial impulse, including a serious effort to locate and build an ethanol plant in the region, in order to reduce the cost of disposing of accumulated fuels. Such a development would increase the feasibility of fuels management and increase the probability of eventually reducing fire danger and damage.

The QLG's proposals were built on the knowledge, creativity, and energy of its members. The group also relied on the credibility and the successes of other groups in the area in designing environmental management and restoration projects—most notably, the Plumas Corporation, the local economic development agency, and Feather River Coordinated Resource Management (CRM), a local collaborative planning group whose members included representatives from state, federal, and local agencies as well as local interest groups (Kusel 1996b). QLG also has drawn freely on the expertise of the staff in public agencies, including the Forest Service and the California Department of Forestry and Fire Protection.

QLG presented the Forest Service with proposals for land management designed to promote ecosystem and local economic sustainability. These plans emphasized the preservation of larger trees, the harvest of small-diameter trees to supply a local mill, fuels management, and the restoration of certain habitats (Terhune 1999, 8). Although the Forest Service remained unreceptive to the QLG's ideas, the group received support from many public officials. QLG made a decision to pursue legislation in Congress that would force the Forest Service to implement the group's recommendations.

National and regional environmental groups were extremely concerned about the legislation and specific aspects of the QLG's proposal. The environmental groups were also concerned about establishing an undesirable precedent that would effectively allow local groups to appropriate the planning and environmental protection responsibilities entrusted to the Forest Service (Duane 1998, 792). Since QLG had formed outside of federal, state, and local government, it was not subject to requirements of public disclosure or accountability. Environmental leaders argued that such arrangements could effectively transfer management authority to local communities. Furthermore, this kind of arrangement could reverse environmental gains won as a result of hard-fought legal battles (McCloskey 1996, 28). Environmental leaders repeatedly expressed concern about the consequences of what they perceived to be devolution of decisionmaking authority from the agency, even though the Forest Service retained responsibility for making resource management decisions in Quincy. They made opposition to the QLG legislation a focal point of their agendas and public campaigns.[21] Although the arguments against the QLG proposal received serious consideration during congressional debate, few members of Congress voted against the bill.[22] Despite substantial opposition from national and regional environmental groups, after a series of delays in Congress, the legislation known as the Herger–Feinstein Quincy Library Group Forest Recovery Act was enacted and signed into law as part of the appropriations legislation in October 1998.

The legislation initiated a separate planning process for the Plumas and Lassen National Forests and the Sierraville Ranger District of the Tahoe National Forest, directing the Forest Service to develop a plan to implement the QLG proposal for a period of five years. A final EIS and Record of Decision were released in August 1999 (U.S. Forest Service 1999). Despite vocal opposition from many environmental groups, the Forest Service approved the QLG pilot project but provided for the possibility of slight modifications to permit additional protection for owl habitat.

The outcome of the QLG's plan is somewhat uncertain at this time. The Forest Service's decision has been appealed on different grounds by parties on different sides of the issue.[23] The Sierra Nevada Conservation Framework will set new guidelines for the conservation of spotted owl habitat and, in turn, will determine whether the specifics of the QLG plan meet the requirements of ESA, NFMA, and NEPA. Additionally, should the U.S. Fish and Wildlife Service determine that the survival of the California spotted owl is in doubt, additional modifications may be required.

The QLG's initiative in policy design, with its attention to environmental issues, local economic concerns, and operational aspects of management prescriptions, remains controversial and experimental. A clear majority of the environmental groups continues to be extremely dissatisfied with the Quincy initiative.[24] Even where environmentalists can and do appreciate specific and limited aspects of the proposals, they regard QLG and its overall objectives with great suspicion, if not outright hostility. Despite QLG's progress thus far, environmental interests continue to challenge the scientific assumptions and the ecological merit of the group's proposals (Edelson 2000).

Since the enactment of QLG legislation, interest and involvement has remained high. QLG and interests revolving around the group are still struggling to advance their own objectives, and the outcome of the process is far from certain. It is worth recalling that QLG and other unconventional approaches to forest planning and management in the region all sprang up after the possibility of a legal challenge caused the Forest Service to reconsider its land management plans resulting in the CASPO-inspired policies introduced in 1993. Despite the QLG legislation, QLG, environmental groups, the Forest Service, and others remain constrained legally and politically by the CASPO policy. In Quincy, the participants and stakeholders are engaged in a still-unfolding "bargaining in the shadow of the law" (Mnookin and Kornhauser 1979, 5, 950), wherein the parties bargain and negotiate agreements on the basis of their understanding of what results the law will sanction. Although the quest for solutions is subject to many constraints, a multitude of opportunities for creative and practical approaches to many aspects of ecosystem management and conservation remain unexplored.

NATIONAL FOREST PLANNING AND MANAGEMENT IN 2000 AND BEYOND

Contemporary Management Setting

Over the past forty years, the Sierra Nevada has undergone remarkable changes. While great expanses of the Sierra Nevada remain relatively wild and undeveloped, the rest of the region has experienced major demographic, economic, and social changes, all of which have implications for resource management (Centers for Water and Wildland Resources 1996, vol. 1, 36–40). Four transformative and interactive forces are reshaping the relationship between people and the natural environment of the region. These forces, discussed in the *Final Report of the Sierra Nevada Ecosystem Project* (Centers for Water and Wildland Resources 1996), are summarized as follows.

- *Continuing population growth and changing patterns of human settlement and development.* Development of several kinds—urban, exurban, commercial, and recreational—affects ecosystem services and has increased the diversity of values and issues that influence environmental policy and governance (Duane 1999).

- *Capitalization of the costs of ecosystem maintenance and environmental risk.* As the region shifts from extractive use of resources, new sources of funding must be found to cover the costs of ecosystem maintenance and restoration. Markets for ecosystem services, however, are largely undeveloped, resulting in a lack of investment in the natural systems in the Sierra Nevada.
- *Governmental coordination and efficiency.* The institutional arrangements for ecosystem management are lacking in at least two areas. First, agency appropriations, designed to support production of timber and other commodities, supply substantially smaller amounts for administration of nonconsumptive uses. Second, intergovernmental and interagency cooperation is reshaping certain aspects of resource administration.
- *Citizen activism and institutional responses.* Residents of the Sierra Nevada and others with extensive knowledge about the region are important sources of human capital. Activism influences, challenges, redirects, and—in some cases—may even replace resource management institutions.

These social, economic, and political forces create a unique mixture of concerns and opportunities in the Sierra Nevada. The Forest Service, despite considerable expertise and technical proficiency, has not been able to respond fully to the effects of these forces. In critical areas, legislative and administrative institutions do not allow the Forest Service sufficient flexibility to account for these forces in policy. In other instances, legal rules and agency tradition have not permitted the kind of initiative or leadership required to encourage cooperation (Centers for Water and Wildland Resources 1996, vol. 2, chap. 20). As a result, the agency generally is not able to take advantage of the dynamism in the region, even where the effect of these forces apparently would support aspects of the Forest Service's mission and objectives.

Rethinking the Forest Service's Approach to Ecosystem Management

Scientific assessment of the natural resources of the Sierra Nevada is central to planning and managing the region's national forests. Scientific knowledge helps the public and decisionmakers understand the issues and choices involved in resource management and stewardship. Although scientific research has played a central role in transforming natural resource policy, improving the state of scientific knowledge and its use in planning is not sufficient to move forward the planning and management of national forest lands in the Sierra Nevada. The report of the California Spotted Owl Federal Advisory Committee (USDA 1997) focused on the deficiencies of the revised draft EIS (U.S. Forest Service 1996). Examining the scientific adequacy of planning proposals is necessary to ensure that management alternatives are consistent with scientific understanding, but a multitude of other factors—social, economic, and institutional—overlay the biological systems of the Sierra Nevada and exert powerful influences on resource management.

Only a portion of the issues plaguing policy development and implementation in the Sierra Nevada region pertain to scientific controversies. Even where differences over science are important components of the disputes, the heart of the argument often is not solely a disagreement over science. It may be a different set of values and priorities, or the basic desire to develop, preserve, or restore a particular landscape, resource, or ecosystem (Noss and Cooperrider 1994). Repeated efforts to resolve outstanding controversies have not ended conflict over the conservation and management of ecosystems and associated resources in the Sierra Nevada. Despite extensive research and far-ranging discussions on the question of scientific and technical improvements to ecosystem stewardship (see Centers for Water and Wildland Resources 1996; California State Board of Forestry 1996), comparatively little attention has been given to the institutional aspects of the problems that resource planning and administration face on public land in the region. Until these issues are addressed, many of the obstacles that have plagued the land management planning processes thus far will remain.

Moving Ahead: Program Development, Planning, and Implementation

Can the current institutional structures for planning and management produce policy that responds to changes in scientific knowledge, ecological conditions, and social concerns? In the past, the Forest Service proposed plans and administered lands and resources for which it had primary responsibility, consistent with the legislative purposes for which the national forests had been established.[25] Nominally, the same arrangement appears to apply today. NFMA calls for the "multiple use and sustained yield" of forest resources. More than a decade of land management planning has brought about a new and very different equilibrium between environmental protection and commodity production than previously existed. The Forest Service is now largely a quasi-regulator of land and resource uses on the national forests, whereas previously it was administrator of those functions. The difference may appear subtle, but it is significant. Sustained yield, with its dual emphases on conservation and use, remains part of the legal standard, but achieving ecological sustainability of forest resources—with little or no emphasis on consumptive resource uses—has become the aspirational objective of national forest management (Dombeck 1998).

Current management concerns are driven by complicated environmental issues, such as threatened and endangered species and cumulative impacts on watersheds. Responses to these issues are determined by laws (for example, NEPA, NFMA, ESA, and the Clean Water Act) and attendant administrative regulations that intentionally restrict agency discretion in certain matters without regard to the impact they may have on other Forest Service missions. As a result, Forest Service decisionmakers often do not exercise authority commensurate with the tasks the agency is called upon to perform. In other cases, the agency does not possess resources or organizational capacity adequate to meet these objectives.

NFMA, NEPA, and ESA have increasingly restricted the Forest Service to applying scientific information to environmental and resource management concerns. Despite the agency's new functional role, legislative and financial support for the practical aspects of this shift have not been commensurate with the change (Centers for Water and Wildland Resources 1996, vol. 1, chap. 3). Theoretically, this approach to managing national forests and conserving resources is valid; practically, however, the legacy of conflicting objectives pertaining to the development and use of natural resource commodities within national forests is extremely problematic. The lack of adequate funding for ecologically sensitive natural resource management restricts the agency in its ability to accomplish its objectives. This failure is especially apparent when existing measures do not address variable local conditions. For example, many of the new prescriptions for fuels management, riparian restoration, and so forth bring in little or no revenue, yet are expensive to undertake. Thus, the implementation of these alternatives depends on budgetary appropriations for agency actions beyond the agency's control.

To its credit, the Forest Service is trying to redress the prior deficiencies in agency decisionmaking and administration (Centers for Water and Wildland Resources 1996, chap. 3). Nevertheless, many current initiatives sponsored by the Forest Service and other groups to foster ecologically sensitive management suggest that a cooperative approach is critical, both to the success of these efforts and to the solution of various national forest policy issues (Dombeck 1998). The gap between the ideal of collaboration and the existing legal and financial structure (which still precludes collaborative decisionmaking) must be considered if the objective of improving the stewardship of the national forests is to be taken seriously.

Accordingly, it has been difficult for the agency to effectively address the ecological issues highlighted by the NEPA process. Unresponsive institutions can neither resolve significant issues concerning national forest management issues nor effectively address underlying ecological concerns. Logically, the organization and institutional structure of the national forests should be examined to determine whether resource management and stewardship functions could be improved. Less-than-optimal institutional arrangements may need to be redesigned or replaced.

The quasi-regulatory approach of today's Forest Service contrasts with the ends-oriented approach that characterized the agency for much of its history (Hays 1969; Clarke and McCool 1984). Society clearly no longer wishes the Forest Service to accomplish the objectives it once did. A planning process with outcomes that cannot be implemented because the institutional structures cannot or will not support them, however, is no solution.[26] If the present policy objective is to improve the development and delivery of ecosystem and resource stewardship, then a more dynamic organizational approach is essential. It would emphasize not only planning and assessment but also accomplishment of the objectives that have been set out. In an organization with a more anticipatory approach to resource management and conservation, decisions would be made only after the implications of various choices had been fully considered. This kind of consideration would evaluate proposals in terms of their scientific implications as well as the likelihood that policy proposals will be successfully implemented and their objectives achieved. The agency's apparent lack of systematic attention to the result is why the ends-oriented approach that local collaborative groups have adopted has become increasingly compelling to many interested parties.

Operational difficulties are magnified by a lack of coordination in many existing programs. A strategic plan to achieve ecological and other goals, carefully thought out and sensibly implemented, may be of considerable value. It might enable more active integration of federal and nonfederal lands in a range of cooperative ventures designed to achieve an entire spectrum of forest-related goals, not only RPA timber targets.

The Forest Service now operates in a context that requires local communities, counties, and state agencies to recognize that they are, in many cases, capable of playing a constructive role in policy innovation and implementation. Almost all current concerns can benefit from area-specific expertise, greater local participation, and regional coordination. Policy choices should be made with sensitivity to local social, economic, and environmental conditions, and they should permit an active collaboration to *facilitate* policy implementation. This approach would ensure that management methods are appropriate to scientific prescriptions and can be implemented locally. Attention to the details of resource conservation will help ensure sound management and ecosystem stewardship.[27]

Recent proposals for reform offered by sources outside the Forest Service are demonstrably more comprehensive than contemporary Forest Service planning.[28] Increasingly, local and other collaborative groups are designing management systems that link scientific knowledge with management prescriptions and the management prescriptions with economic feasibility, community well-being, and other socioeconomic factors (Kusel and others 1996). The specific scientific validity of these ideas and individual proposals, like the QLG proposals, must be reviewed individually. The virtues of such an approach, however, particularly in its pragmatic outlook, should not be overlooked.

Environmental interests are skeptical of the rationale and the impetus for community-based resource management. Throughout the planning process in the Sierra Nevada, environmentalists fought for more complete analysis of the impacts of proposed plans with the public and local groups by the Forest Service. Local and regional environmental groups have worked closely with local environmentalists on conservation issues. These groups have fought arduous battles over forest planning decisions. Justifiably, they do not want to see the results of these victories eroded, nor do they wish to return to an earlier era of resource management that favored the interests of local communities over sound conservation practices. They also are concerned that if planning and decisionmaking devolve to a subregional or local level, planning processes will occur in settings in which environmental perspectives are not sufficiently valued (McCloskey 1996). Where environmentalists are underrepresented, outmaneuvered, or pressured to make inappropriate compromises, the result may be environmentally undesirable. Finally, relaxing or changing the legal mandates now in place to suit local conditions is also troubling, because it can further dilute the influence of environmental interests, especially in ensuring that the resource prescriptions adopted are ecologically sound (Blumberg 1998).

For an ecological approach to natural resource management to succeed, community-based resource management is probably not required, nor is it sufficient by itself. Initiatives like those of QLG do not please everyone, but they are part of the policy landscape in the Sierra Nevada. For lasting progress to be made, acrimonious disputes over natural resources policy, use, and management must be left behind to pursue positive outcomes. The diverse and creative community of actors in ecosystem and resource policy—concerned residents and other citizens, agency planners, managers, public officials, organizations at all levels of government, businesses, and nongovernmental organizations—all have something of value to contribute. Citizens, environmental interests, and others are exploring and experimenting with new methods of participation, outreach, and policy development in resource management. This energy will help create or recreate institutions to implement programs and achieve the desired results. The results they accomplish will determine the future management objectives for the national forests in the Sierra Nevada.

LESSONS FOR THE FUTURE

Examining forty years of forest policy allows for reflection on the complexity of the institutional setting in which these policies operate. Understanding the sources and contours of past conflicts over resource management and conservation cannot itself provide a solution to future conflicts, but insights gained from the experience provide the means to address continuing controversies in environmental management. The reversals in Forest Service planning over the last several decades reflect the impact that the changing values of American society have had on public policy. Environmental and natural resource–related policies previously had operated without explicitly considering the ecosystem as a point of reference for policy formation, implementation, or evaluation, but changing public values and priorities forced the agencies to do so. Efforts to apply the knowledge gained as a result of SNEP and other studies are now the central focus of ecosystem planning in the region. As the Forest Service's experience with planning has demonstrated, better collection and analysis of natural and socioeconomic data are necessary to ensure that policies will be legally adequate, publicly accepted, and successfully implemented. A great deal of information and experience gained as a result of successive statutory and administrative reforms resides in the public, agency officials, and scholars. Whether this knowledge will enable better environmental management and translate into more stable policy for the Sierra Nevada remains to be seen.

The source of direction for national forest management also has changed subtly but surely. The centralization and standardized approach to resource management employed by the Forest Service under the auspices of MUSYA and in the earlier days of NFMA has been replaced (Dombeck 1998). Even though NFMA continues in effect, the Forest Service no longer controls national forest policy. Instead, mandatory provisions of the law and regulations, such as the regulatory requirement to provide for "minimum viable populations," mean that the regional and local landscapes, watersheds, and their resources are now the focus of attention. Assessments of the viability of these resources come first, and the results of these assessments now directly influence planning for the future.

The environmental goals embodied in current federal environmental law and natural resource policy emerged as a result of social and economic changes. Organizational structures and financing mechanisms for resource conservation and management reflect older institutional objectives, however; they have not undergone substantial change and must struggle to meet the demands of new policy objectives. The inability of the Forest Service and other public agencies to interpret and respond effectively to the public's priorities regarding national forest management may be changing. Nevertheless, these agencies lack the institutional capacity or authority to fully develop and implement ecosystem conservation agendas and resource management programs. And while the frontier of scientific knowledge has advanced in light of study, research, and experimentation, additional investigations—followed by institutional reform, experimentation, and redesign—could support advances in scientific understanding and complete a transformation in national forest policy for the Sierra Nevada.

ACKNOWLEDGMENTS

The author wishes to convey his gratitude to all of the individuals who reviewed and commented on earlier versions of this research. They provided invaluable and constructive criticism.

NOTES

[1]The Forest Service administers the Lassen, Plumas, Tahoe, Eldorado, Stanislaus, Sierra, Inyo, and Sequoia National Forests and the Lake Tahoe Basin Management Unit as part of the Pacific Southwest Region of the National Forest System. Sections of the Humboldt–Toiyabie National Forest are in the Sierra Nevada but administered as part of the Intermountain Region.

[2]McConnell (1966) argued that the doctrine of multiple use effectively gave the agency discretion to implement its own policies with little regard to public opinion.

[3]As early as 1929, at the instigation of foresters within the agency, timber harvesting and other management activities already had been limited in certain areas within the national forests.

[4]Among the problems identified with selection logging is "highgrading." This refers to the effect of repeated selection and harvest of the best trees. The result is that the largest, most vigorous trees of a stand are removed and not allowed to reproduce, diminishing the genetic quality and commercial value of the trees over time (see Nyland 1996).

[5]The case also involved the Multiple Use Sustained Yield Act of 1960. The court held that this statute did not amend the timber harvest provisions of the 1897 Organic Act.

[6]Sample (1990) argued that the politics of the budgetary process regarding Forest Service budgets have long diverted it from any great reliance on RPA projections and budgets.

[7]NFMA allowed the agency to use clearcutting and even-aged management, but made such use subject to several restrictions intended to protect the forest landscape. NFMA authorized the use of clearcutting in stands that included "immature" trees in national forests, but only after completion of comprehensive land and resource planning that demonstrated the efficacy of applying the technique to particular forest stands.

[8]See http://www.r5.fs.fed.us/sncf/framework/design_paper/design_paper_1.4.html.

[9]Related silvicultural methods included seed-tree cutting and overstory removal. Even though these methods were not technically clearcutting, often they are criticized because their effects are similar to those of clearcutting, namely, the removal and replacement of entire stands of timber. It is worth noting that the final EIS did acknowledge that uneven-age management might succeed if properly conducted. In the Plumas, the preferred alternative was revised to manage 800 acres per year under an uneven-age silvicultural system for purposes of research.

[10]All planning data are public information, freely obtainable for examination in detail, although agency reluctance occasionally made this difficult. Some critics visited Forest Service offices and delved deeply into the data as part of their examination of the draft EIS and the draft land management plans. Local environmental activists and industry representatives (often with the assistance of consultants) studied voluminous FORPLAN runs and other data.

[11]On the Tahoe National Forest, Forest Service proposals in the draft plan called for the extensive use of clearcutting and the harvest of timber in former roadless areas. The ensuing controversy resulted in more than 12,000 letters to the Forest Service, supporting and opposing various aspects of the land management plan.

[12]Data from national forest timber sales reflected this increase (Verner and others 1992, 240–1).

[13]In a January 19, 1993, letter to Forest Service Chief Dale Robertson, Congressman George Miller, Chairman of House Committee of Natural Resources, and seven other members of Congress called for the creation of a panel of scientists who would produce a map of old-growth ecosystems in the Sierra Nevada and prepare a report to Congress that was to include a range of alternatives for Sierra Nevada management (see Sierra Biodiversity Institute's Web page at http://www.oro.net/~sbihome/history2.htm#Establishment [accessed April 13, 2000]).

[14]See also http://ceres.ca.gov/snep/pubs/ (accessed April 11, 2000).

[15]Certain practices associated with logging had been observed to increase fire risk (see Centers for Water and Wildland Resources 1996, vol. 2, 1173).

[16]The "sensitive species" included other furbearing mammals, such as the fisher (*Martes penannti*) and the Pacific marten (*Martes americana*).

[17]Sierra Nevada EIS Timeline, http://www.r5.fs.fed.us/sncf/eis/eis_timeline. html (accessed March 6, 2000).

[18]The spotted owl or another plant or animal species in the Sierra Nevada, if listed as a "threatened" or "endangered" species under the Endangered Species Act, would force agencies to reconsider land and resource uses to ensure that any activities are not injurious to the viability of the species.

[19]See the Quincy Library Group Community Stability Proposal, http://www.qlg.org/pub/agree/comstab.htm (accessed April 14, 2000).

[20]As originally enacted, the National Forest Revenue Act of 1908 required the federal government to give 5% of revenues from national forests to the counties in which the forests were located. In 1913, the percentage was increased to the current level of 25% of national forest revenues.

[21]See the diversity of perspectives on the Quincy Library Group at http://www.qlg.org/pub/contents/perspectives.htm (accessed April 14, 2000).

[22]The original legislation passed overwhelmingly in the U.S. House of Representatives. In the Senate, legislative "holds" prevented the legislation from coming to a vote. Eventually, the QLG bill was incorporated into the 1999 Department of the Interior and Related Agencies Appropriations Act.

[23]See *Appeal by the Quincy Library Group of the Final EIS and Record of Decision for the Herger–Feinstein Quincy Library Group Forest Recovery Act*, November 4, 1999, http://www.qlg.org/pub/act/appeal.htm (accessed March 5, 2000) and Sierra Nevada Forest Protection Campaign, *Conservationists Appeal Quincy Decision*, http://www.sierraforests.org/html/updates101899.html (accessed October 18, 1999).

[24]See, for example, Sierra Nevada Forest Protection Program, http://www.sierraforests.org/index.html (accessed April 14, 2000).

[25] In *United States v. New Mexico*, 438 U.S. 696, 707 n.14 (1978), the majority opinion noted that "close examination of the language of the Act, however, reveals that Congress only intended national forests to be established for two purposes. Forests would be created only 'to improve and protect the forest within the boundaries,' or, in other words, 'for the purpose of securing favorable conditions of water flows, and to furnish a continuous supply of timber.'"

[26]See, for example, the Committee of Scientists' *Final Report*, at http://www.fs.fed.us/news/science/cos-ch1pt2.pdf (accessed December 21, 1999).

[27] The "hazard fuels management" program is one example of this approach. Supported by the Clinton administration, a coalition of public agencies, and other interests, including the Wilderness Society, it seeks to increase funding for prescribed burning and other means of returning fire to a more natural role in the landscape, although it is not tied to a particular locale.

[28]See Committee of Scientists' *Final Report,* at http://www.fs.fed.us/news/science/cos-ch1pt2.pdf (accessed December 21, 1999); see also Thoreau Institute, *Public Land Research and Analyses Second Century Report,* http://www.ti.org/2c.html (accessed March 5, 2000).

REFERENCES

Beckwitt, Steve, Eric Beckwitt, and Willow Beckwitt. 1986. Interview with Steve, Eric, and Willow Beckwitt, Sierra Club, Sierra Nevada Group, Timber Issues Task Force, Nevada City, California (June 15).

Behan, Richard. 1981. RPA/NFMA—Time to Punt. *Journal of Forestry* 79:802, 805.

Bernard, Ted, and Jora Young. 1997. *The Ecology of Hope.* East Haven, Connecticut: New Society Publishers.

Blumberg, Louis. 1998. *An Environmentalist's Commentary on the Quincy Library Group.* Address at the University of California at Berkeley (February 10).

California State Board of Forestry. 1996. *California Fire Plan* (March).

Centers for Water and Wildland Resources. 1996. *Status of the Sierra Nevada, Final Report to Congress.* Sierra Nevada Ecosystem Project, University of California at Davis. Davis: Centers for Water and Wildland Resources.

Clarke, Jeanne Nienaber, and Daniel McCool. 1984. *Staking Out the Terrain.* Albany: State University of New York Press.

Clary, David. 1985. *Timber and the Forest Service.* Lawrence, Kansas: University Press of Kansas.

Clawson, Marion. 1983. *The Federal Lands Revisited.* Washington, D.C.: Johns Hopkins University Press for Resources for the Future.

Craine, Jim. 1986. Interview with Jim Craine, Vice President of the California Forestry Association (formerly the Western Timber Association) (June).

Culhane, Paul. 1981. *Public Lands Politics: Interest Group Influence on the Forest Service and the Bureau of Land Management.* Baltimore: Johns Hopkins University Press for Resources for the Future.

Dana, Samuel T., and Sally K. Fairfax. 1980. *Forest and Range Policy,* 2nd ed. New York: McGraw-Hill.

Doak, Sam C., and Jonathan Kusel. 1996. Well-Being in Forest-Dependent Communities, Part II: A Social Assessment Focus. In *Status of the Sierra Nevada, Final Report to Congress,* by Centers for Water and Wildland Resources, Sierra Nevada Ecosystem Project, University of California at Davis. Davis: Centers for Water and Wildland Resources, vol. 2, 375–402.

Dombeck, Michael. 1998. *A Gradual Unfolding of a National Purpose: A Natural Resource Agenda for the 21st Century.* Address to Forest Service Employees (March 2).

Dowdle, Barney, and Steve Hanke. 1985. Public Timber Policy and the Wood Products Industry. In *Forestlands: Public and Private,* edited by Robert T. Deacon and M. Bruce Johnson. San Francisco, California: Pacific Institute for Public Policy Research, and Cambridge, Massachusetts: Ballinger Publishing Co.

Duane, Timothy. 1998. Community Participation in Ecosystem Management. *Ecology Law Quarterly* 24: 771–797.

———. 1999. *Shaping the Sierra: Nature, Culture, and Conflict in the Changing West.* Berkeley: University of California Press.

Ecology Law Quarterly. 1972. Mineral King: A Case Study in Forest Service Decision Making. *Ecology Law Quarterly* 2: 493.

———. 1976. Mineral King Goes Downhill. *Ecology Law Quarterly* 5: 555.

Edelson, David. 1997. Address at University of California at Berkeley (February).

———. 2000. Interview with David Edelson, Natural Resources Defense Council, San Francisco, California (January 25).

Ewing, Robert. 1992. Interview with Robert Ewing, Director, Forest and Rangeland Assessment Program, California Department of Forestry and Fire Protection (December 18).

Fisk, Deborah L., and others. 1997. Mediated Settlement Agreement for Sequoia National Forest. In *B. Giant Sequoia, an Evaluation, Status of the Sierra Nevada, Final Report to Congress,* by Centers for Water and Wildland Resources, University of California at Davis, Report No. 40, 277 (March).

Forest and Rangeland Assessment Program. 1988. *California Department of Forestry and Fire Protection, California's Forests and Rangelands: Growing Conflict over Growing Uses.* California Department of Forestry and Fire Protection.

Foss, Philip. 1960. *Politics and Grass: The Administration of Grazing on the Public Domain.* Seattle: University of Washington Press.

Franklin, Jerry F., and Jo-Ann Fites-Kauffman. 1996. Assessment of Late-Successional Forests of the Sierra Nevada. In *Status of the Sierra Nevada, Final Report to Congress,* by Centers for Water and Wildland Resources, Sierra Nevada Ecosystem Project, University of California at Davis. Davis: Centers for Water and Wildland Resources, vol. 2, chap. 21.

Franklin, Jerry F., and others. 1996. Alternative Approaches to Conservation of Late-Successional Forests in the Sierra Nevada and Their Evaluation. In *Status of the Sierra Nevada, Final Report to Congress,* by Centers for Water and Wildland Resources, Sierra Nevada Ecosystem Project, University of California at Davis. Davis: Centers for Water and Wildland Resources, vol. 2, chap. 3.

Friedmann, John. 1987. *Planning in the Public Domain: From Knowledge to Action.* Princeton, New Jersey: Princeton University Press.

Handler, Joel. 1988. Dependent People, the State, and the Modern/Post Modern Search for the Dialogic Community. *UCLA Law Review* 35:999.

Hays, Samuel P. 1969. *Conservation and the Gospel of Efficiency: The Progressive Conservation Movement 1890–1920.* New York: Atheneum.

Hirt, Paul. 1994. *A Conspiracy of Optimism: Management of the National Forests since World War Two.* Lincoln: University of Nebraska Press.

Jackson, Michael, and Tom Nelson. 1996. Address at University of California at Berkeley (March 21).

Knudson, Tom. 1999. Groups to Urge Legal Shield for California Spotted Owl. *The Sacramento Bee* (June 23).

Kusel, Jonathan. 1996a. Well-Being in Forest-Dependent Communities, Part I: A New Approach. In *Status of the Sierra Nevada, Final Report to Congress,* by Sierra Nevada Ecosystem Project, University of California at Davis. Davis: Centers for Water and Wildland Resources, vol. 2, 361–74.

Kusel, Jonathan. 1996b. Coordinated Resource Management. In *Status of the Sierra Nevada, Final Report to Congress,* by Sierra Nevada Ecosystem Project, University of California at Davis. Davis: Centers for Water and Wildland Resources, vol. 3, chap. 24.

Kusel, Jonathan, and others. 1996. The Role of the Public in Adaptive Ecosystem Management. In *Status of the Sierra Nevada, Final Report to Congress,* by Sierra Nevada Ecosystem Project, University of California at Davis. Davis: Centers for Water and Wildland Resources, vol. 2, 611–24.

Leisz, Doug. 1995. Interview with Doug Leisz, Deputy Chief, United States Forest Service (retired) (September).

Marston, Ed. 1997. The Timber Wars Evolve into a Divisive Attempt at Peace. *High Country News* (September 29).

Martin, Glen. 2000. A Petition to Help California Bird: Endangered Status Sought for Kin of Northern Spotted Owl—Only 2,000 Left. *San Francisco Chronicle* (April 14).

McCloskey, Michael. 1996. The Skeptic: Collaboration Has Its Limits. *High Country News* 28(May 13).

McCloskey, Michael J. 1966. The Wilderness Act of 1964: Its Background and Meaning. *Oregon Law Review* 45: 288.

McConnell, Grant. 1966. *Private Power and American Democracy.* New York: Knopf.

Mnookin, Robert H., and Lewis Kornhauser. 1979. Bargaining in the Shadow of the Law. *Yale Law Journal* 88: 5, 950.

Muir, John. 1961. *The Mountains of California.* Garden City, New York: Doubleday.

National Forest Service. 1907. *The Use Book.* U.S. Department of Agriculture, 16.

Nijhuis, Michelle. 1999. Flagstaff Searches for Its Forest Future. *High Country News* 8–12 (March 1).

Noss, Reed F., and Allen Cooperrider. 1994. *Saving Nature's Legacy, Protecting and Restoring Biological Diversity.* Washington, D.C.: Island Press.

NRDC (Natural Resources Defense Council). 1991. Appeal of the Tahoe National Forest Land Management Plan (March 15).

———. 1993. Appeal to F. Dale Robertson Regarding his January 13, 1993 Decision Notice (March 1, 1993) (on file with the author).

Nyland, Ralph. 1996. *Silviculture: Concepts and Applications.* New York: McGraw-Hill.

O'Toole, Randal. 1988. *Reforming the Forest Service.* Washington, D.C.: Island Press.

Reich, Robert. 1985. Public Administration and Public Deliberation: An Interpretive Essay. *Yale Law Journal* 94:1617–40.

Rey, Mark. 1991. Address at the American Forest Resource Alliance, Berkeley, California (March 14).

Robertson, F. Dale. 1992. Memorandum from F. Dale Robertson, Chief of the U.S. Forest Service, on ecosystem management (July 19) (on file with the author).

Rosenbaum, Kenneth. 1984. Forest Planning—Bound for the Courts Again. *Environmental Law Report* 14 (May):10195.

Ruth, L., and R. Standiford. 1994. *Conserving the California Spotted Owl: Impacts of Interim Policies and Implications for the Long Term.* Report of the Policy Implementation Planning Team to the Steering Committee for the California Spotted Owl Assessment, Wildland Resource Center, University of California at Davis (May).

Sample, V. Alaric. 1990. *The Impact of the Federal Budget Process on National Forest Planning.* New York: Greenwood Press.

San Francisco Chronicle. 1996. Recreation's Growing Impact. D8 (September 19).

San Francisco Chronicle and Examiner. 1999. Calls for Controlled Burns Resisted as Smog, Forest Goals Clash. April 12.

Sedjo, Roger. 1999. Mission Impossible. *Journal of Forestry* 97(May): 5, 13–14.

Solnit, Rebecca. 1997. Among the Giants: California's Sequoias May Be More than 3,000 Years Old, but They're Running Out of Time. *Sierra* (July/August): 30–7, 63.

Southwest Center for Biological Diversity and Sierra Nevada Forest Protection Campaign. 1999. *A Preliminary Report on the Status of the California Spotted Owl in the Sierra Nevada.* Unpublished memorandum (June 21).

Stewart, Richard B. 1975. The Reformation of American Administrative Law. *Harvard Law Review* 88: 1669.

Stewart, Ronald E., and others. 1992. Giant Sequoia Management in the National Forests of California. In *Symposium on Giant Sequoias: Their Place in the Ecosystem and Society,* U.S. Department of Agriculture, U.S. Forest Service, Pacific Southwest Research Station, Visalia, California (June 23–25).

Stewart, William. 1996. Economic Assessment of the Ecosystem. In *Status of the Sierra Nevada, Final Report to Congress,* by Sierra Nevada Ecosystem Project, University of California at Davis. Davis: Centers for Water and Wildland Resources, vol. 3, chap. 38.

Stroup, Richard, and John Baden. 1983. *Natural Resources: Bureaucratic Myths and Environmental Management.* San Francisco: Pacific Institute for Public Policy Research, and Cambridge, Massachusetts: Ballinger.

Taylor, Serge. 1984. *Making Bureaucracies Think.* Stanford, Calififornia: Stanford University Press.

Teeguarden, Dennis. 2000. Interview with Professor Dennis Teeguarden, Berkeley, California (February 1).

Terhune, George. 1999. *The Quincy Library Group, a Case Study.* CDF Discussion Paper. Gainesville, Florida: Conservation and Development Forum.

Thomas, Jack Ward, and others. 1990. *A Conservation Strategy for the Northern Spotted Owl.* Report. Portland, Oregon: Interagency Scientific Committee to Address the Conservation of the Northern Spotted Owl.

USDA (U.S. Department of Agriculture). 1997. *Final Report of the California Spotted Owl Federal Advisory Committee,* 2-1–2-9.

U.S. Forest Service. 1960. *Report of the Chief.* Washington, D.C.: U.S. Department of Agriculture, 19.

———. 1972. *Forest Service Manual.* Washington, D.C.: U.S. Department of Agriculture, sect. 8213.

———. 1976. *Environmental Analysis and Alternatives for the Northern California Planning Area Guide.* 5.

———. 1988. *Plumas National Forest FEIS for the Management Plan.* Pacific Southwest Region.

———. 1990a. *Mediated Settlement Agreement for Sequoia National Forest, B. Giant Sequoia.* Pacific Southwest Region.

———. 1990b. *Tahoe National Forest Land Management Plan and Environmental Impact Statement.* Pacific Southwest Region.

———. 1990c. *Tahoe National Forest FEIS for the Management Plan.* III-26.

———. 1993a. *California Spotted Owl Sierran Province Interim Guidelines.* Pacific Southwest Region. III-1-2 (January).

———. 1993b. *California Spotted Owl Sierran Province Interim Guidelines Environmental Assessment.* Pacific Southwest Region. III-1-2 (January).

———. 1993c. *Decision Notice and Finding of No Significant Impact for California Spotted Owl Sierran Province Interim Guidelines.* Pacific Southwest Region. DN-13-15 (January).

———. 1994. *General Technical Report.* Pacific Southwest Research Station. PSW GTR-151, 154–5.

———. 1995. *Managing California Spotted Owl Habitat in the Sierra Nevada National Forests of California: An Ecosystem Approach.* Pacific Southwest Region. Draft Environmental Impact Statement (January).

———. 1996. *Managing California Spotted Owl Habitat in the Sierra Nevada National Forests of California: An Ecosystem Approach.* Pacific Southwest Region. Revised Draft Environmental Impact Statement (August) (on file with author).

———. 1998a. Memorandum from the USDA Forest Service, Sierra Nevada Conservation Framework (January 22, 1998) (on file with author).

———. 1998b. Pacific Southwest Region. http://www.fs.fed.us/land/fm/salefact/salefact.htm (accessed February 25).

———. 1999. *Final Environmental Impact Statement: Herger–Feinstein Quincy Library Group Forest Recovery Act* (August 20).

U.S. ORRRC (Outdoor Recreation Resources Review Commission). 1962. *Outdoor Recreation for America: A Report to the President and to the Congress.* Washington, D.C.: ORRRC.

Verner, Jared. 1993. Interview with Jared Verner, Project Leader, Wildlife Monitoring and Range Research, Pacific Southwest Research Station (July 22).

Verner, Jared, and others. 1992. *The California Spotted Owl: A Technical Assessment of Its Current Status.* U.S. Department of Agriculture, U.S. Forest Service (July 15).

Wilkinson, Charles F., and H. Michael Anderson. 1985. *Land and Resource Planning in the National Forests*. Washington, D.C.: Island Press.

Wilson, Carl. 1978. Land Management Planning Processes of the Forest Service. *Environmental Law* 8:461.

Wilson, James. 1947. Letter from Secretary of Agriculture James Wilson to Gifford Pinchot. In *Breaking New Ground,* by Gifford Pinchot. New York: Harcourt, Brace.

Wondolleck, Julia. 1988. *Public Lands Conflict and Resolution: Managing National Forest Disputes.* New York: Plenum Press.

Yaffee, Steven L. 1994. *The Wisdom of the Spotted Owl: Policy Lessons for a New Century.* Washington, D.C.: Island Press.

Zivnuska, John. 1993. Interview with John Zivnuska, Professor and Dean Emeritus, School of Forestry, University of California at Berkeley (March 9).

Index